# Vygotsky and Science Education

Colette Murphy

# Vygotsky and Science Education

Springer

Colette Murphy
School of Education
Trinity College Dublin
Dublin 2, Dublin, Ireland

ISBN 978-3-031-05243-9 ISBN 978-3-031-05244-6 (eBook)
https://doi.org/10.1007/978-3-031-05244-6



This Springer imprint is published by the registered company Springer Nature Switzerland AG
The registered company address is: Gewerbestrasse 11, 6330 Cham, Switzerland

# Contents

# Part I
# Introduction

# Chapter 1
# Introduction

*It is through others, we become ourselves.* **(Vygotsky, 1997, p. 105)**

Lev Semenovich Vygotsky (1896–1934) was widely acclaimed as a scientific genius from Russia, so much so that after his death his brain, along with those of Vladimir Lenin (1870–1924) and Ivan Pavlov (1849–1936), was studied by Russian scientists researching the neuroanatomical basis of exceptional talent. More recently, in the West, Stephen Toulmin (1978) referred to Vygotsky as the "Mozart of Psychology" because his work was prodigious and novel, and has immense contemporary relevance, despite his very early death from tuberculosis (TB) at the age of 37.

## 1.1 Vygotsky and Science Education

This book introduces Vygotsky and his work, emphasising implications for science education. It aims to provide a reference/handbook for university teachers in education and, more specifically, in science education. Researchers in science education can use this text as an introduction to Vygotsky's substantial contribution


C. Murphy, *Vygotsky and Science Education*,
https://doi.org/10.1007/978-3-031-05244-6_1

to theory and practice in the field. A unique contribution in this book is to infuse my experiences of working from a Vygotskian perspective as a school science teacher, a university teacher of both science and science education, and a researcher in science education. My key motivation for writing the book came from highly positive student and colleague responses to the Vygotskian-inspired methodology I used both with students and colleagues. I was also encouraged to write the book by members of the Vygotsky family, when attending Vygotsky summer school immersion programmes in Russia, as a member of the International Vygotsky Society.

I hope that you enjoy reading about science learning, teaching and research from a Vygotskian perspective, which promotes social, as opposed to individual, learning of science. His concepts and research findings have led to some bold ideas in practice, such as:

- Emancipating learners, by harnessing their interests and creating learning environments from them: they lead the learning.
- Engaging learners, by introducing them to some of the greatest and most beautiful achievements in science, be it stories, trips, videos, podcasts, etc. Vygotsky motivated his students to learn by showing, reading, or experiencing in other ways, some of the greatest products of the human soul in art, literature, science, music and other areas of interest that generate strong emotion and lead to the desire to learn more.
- Teaching them a little beyond their current knowledge, with support: whatever they can do today with help, tomorrow they can do it by themselves.

Vygotskian theory is becoming increasingly relevant and important in science learning and teaching in the twenty-first century. Vygotsky's focus on learning as a social process, as opposed to a purely individual one, resonates with current practice in science and science education, for example, co-operation and collaboration. Twenty-first-century learning seeks for students to develop skills relating to different ways of working from those of the twentieth century, to help students develop a global perspective through which people collaborate across geographical and social borders to address major issues, such as climate change and antibiotic resistance.

One of the key aspects of Vygotsky's legacy to science education is his cultural-historical theory (CHT), which proposes that the *social setting* is fundamental to learning in childhood and adulthood and determines the quality of learning that takes place. Learning occurs as an individual engages in social settings, from babyhood to adult learning. Other Vygotskian concepts and practices which are key to science learning include the zone of proximal development (ZPD), concept development, and the importance of imagination and play in learning and in 'doing' science. Vygotsky rejected the objective approach to learning used by many psychologists who established their ideas on what may be considered as an 'average' child, based on specific measurements. Vygotsky maintained that the environment for learning should be attuned to learners in ways that encourage them to learn more for the sake of learning, as opposed to learning only for external motivation (exams, careers, etc.).

The book aims to develop a framework of Vygotskian theory for research and practice in science education, accessible to students, teachers, and researchers in science education for early years, school, college and university, and adult science learning and teaching. Thus, play and imagination are explored in depth in the chapter on science in early childhood learning (Chap. 5); the ZPD is considered in depth in the primary school science chapter (Chap. 6); and concept development in the secondary-level chapter (Chap. 7). Chapters on informal science learning (Chap. 8); higher education science learning and teaching (Chap. 9); and science teacher education (Chap. 10) draw on cultural-historical activity theory (CHAT) the ZPD, imagination and play in science learning, and concept development.

Formal science education currently focuses on curricula and assessment, guided by global competition in terms of international projects such as the Programme for International Student Assessment (PISA) and Trends in International Mathematics and Science Studies (TIMSS). These projects produce rankings of science learning in different countries. Science education could be focused more on solving problems relating to global issues, which might result in earlier and better change.

## 1.2 Structure of This Book

This book has an introduction (Part I), two main content sections (Parts II and III), and ends with an epilogue (Part IV).

Part I: Introduction

Chapter 1: Introduction – provides an overview of this book, which introduces some Vygotskian ideas that are considered useful in the learning and teaching of science at all levels, in both formal and informal contexts. The chapter summarises each of the subsequent chapters and suggests ways to navigate the book (see Table 1.1 below).

Part II: Vygotsky and His Legacy to Science Education

There are three chapters:

Chapter 2: Vygotsky: Life and Legacy – presents a short biography of Vygotsky, including the social and cultural influences on his work. I learned a great deal about Vygotsky's theory and practice by reading and writing about his short life. This chapter includes reference to philosophers, psychologists, scientists, playwrights, poets and novelists whose work influenced Vygotsky's ideas, research and practice.

Chapter 3: Key Vygotskian Concepts for Science Education – introduces key Vygotskian concepts and practices that are relevant to science education, although some concepts are attributed to Vygotsky, but not developed purely by him. This chapter includes concepts such as: ZPD; concept development, CHT; and Vygotsky's early ideas for the development of CHAT, together with his colleague A. N. Leontiev (1981) and later to be more fully developed by Yrjö *Engeström, about 50 years after Vygotsky's death* (Engeström, 1987). CHAT helps us to understand and analyse the relationship between human minds (what we think and feel) and activity (what we

**Table 1.1** Useful cross-referencing for specific science education contexts

| Science education contexts | Cross references | |
|---|---|---|
| | Key chapter | Some additional cross-references from other chapters |
| Science Education Researchers | Chap. 4 | Cross-reference sections: 2.3; 2.4; 3.2; 3.3; 3.4; 3.5; 5.6; 6.5; 6.6; Table 7.1; 7.3; 7.7; 8.2; 9.7; 10.4. |
| Early Years Science Education | Chap. 5 | Cross-reference sections: 3.2; 3.4. |
| Primary School Science Education | Chap. 6 | Cross-reference sections: 3.2; Figure 3.1; 5.3; 5.5; 5.6. |
| Secondary-Level Science Education | Chap. 7 | Cross-reference sections: 3.2; 3.3. |
| Informal Science Education | Chap. 8 | Cross-reference sections: 3.4; Figure 3.2. |
| Science in Higher Education | Chap. 9 | Cross-reference sections: 3.2; 7.3; 7.5; 7.6; 7.7. |
| Science Teacher Education | Chap. 10 | Cross-reference sections: 3.2; 7.5; 7.6; 7.7. |

do). CHAT has attracted a growing interest among academics worldwide since the 1990s. The concept of social constructivism was *not* developed by Vygotsky; it has been attributed erroneously to him, probably due to his CHT of learning, which emphasises social, as opposed to individual learning.

Chapter 4: Vygotsky and Science Education Research – describes Vygotsky's own methodologies and how aspects of his work have more recently been applied in science education research.

Part III: Application of Vygotskian Theory in Science Education Contexts

There are six chapters that highlight Vygotskian ideas and practices in science education in various settings, ranging from early years science learning and teaching to higher education science.

Chapter 5: Vygotsky and Science Learning in the Early Years – describes theory and practice of science learning in pre-school and early primary school. It also introduces some of Vygotsky and other scholars' research on child development.

Chapter 6: Vygotsky and Primary School Science – illustrates ways in which Vygotskian theory and practice have been applied to science learning for children between 4 and 12 years of age. There is a focus on the ZPD.

Chapter 7: Vygotsky and Secondary-Level School Science – focuses on scientific concept development. Much of the traditional learning of science concepts (SCs) has been committing definitions and formulae to memory, which does not always lead to true understanding. This chapter describes some of the Vygotskian approaches which have been shown to improve students' understanding and use of true scientific concepts.

Chapter 8: Vygotsky and Informal Science Learning – introduces a dialectical interrelationship between informal and formal science learning, in which the former provides students with many opportunities to *talk about* science with their fellow students, scientists and science teachers, thus helping students to make sense of their learning.

Chapter 9: Vygotsky and Science in Higher Education – illustrates ways that social constructivist approaches, some based on Vygotskian theory, have been used to help students to develop scientific *habits of mind*, which are required for becoming good scientists (Gauld, 1982).

Chapter 10: Vygotsky and Science Teacher Education – includes initial teacher education (ITE), continuing professional development (CPD) for teachers, and informal science education contexts.

Part IV: Epilogue

The epilogue, presented as Chap. 11, presents a skeleton framework for the application of Vygotskian principles and practice in science education. It also considers the potential influence that Vygotsky's work may have on improving ways of 'doing' science. This is timely, as there are currently many issues in science that require more attention to bridging the gap between the scientific institutions and the world that surrounds them.

## 1.3 Navigating the Book Chapters

This book provides a comprehensive, yet concise, overview of Vygotsky's legacy to science education in many different learning, teaching and research contexts. In Table 1.1 below, I have indicated cross-references to different chapter sections for each context. There is some repetition of Vygotskian concepts throughout the chapters, but this has been kept to a minimum via Table 1.1, which should save readers a lot of time in trying to find useful cross-references between the various chapters.

My first introduction to Vygotsky was while completing my master's thesis on scientific concept development. More experience of Vygotsky began when I started lecturing in learning theories to pre-service teachers, in-service teachers, and master's and doctoral students. I became hooked on Vygotsky's life and legacy to science and science education. I use Vygotskian concepts and practices in my own teaching, learning and research. The positive responses from students ignited my motivation to start this book.

The book is written as a guide for readers who wish to learn more about Vygotsky, and how his inspirational work relates to many different science education contexts, most especially in twenty-first-century learning and teaching. Also, Vygotskian scholars reading this book might be interested in the different ways in which his work has been appropriated in science education.

I hope that you, the reader, will gain some useful ideas, and perhaps inspiration, for learning, teaching and 'doing' science!

## References

Engeström, Y. (1987). *Learning by expanding: An activity-theoretical approach to developmental research*. Orienta-Konsultit.

Gauld, C. F. (1982). The scientific attitude and science education: A critical reappraisal. *Science Education, 66*, 109–121. Retrieved from https://doi.org/10.1002/sce.3730660113

Leontiev, A. N. (1981). *Problems of the development of the mind*. Progress Publishers (Original work published in 1959).

Toulmin, S. (1978, September 28). The Mozart of psychology. *The New York Review of Books*, pp. 51–57.

Vygotsky, L. S. (1997). *The collected works of L. S. Vygotsky, vol. 4. The history of the development of higher mental functions* (R. W. Reiber, Ed.; M. J. Hall, Trans.). Plenum (Original work published in 1931).

# Part II
# Vygotsky and His Legacy to Science Education

# Chapter 2
# Vygotsky – Life and Legacy

## 2.1 Introduction

Lev Semenovich Vygotsky (1896–1934) was widely acclaimed as a scientific genius from Russia, whose work is now highly influential globally in all aspects of education. His focus on social (as opposed to individual) learning has provided a basis for successful collaboration to promote better learning. Vygotsky's work is key to recent developments in science education that have embraced social constructivist approaches to learning and teaching science.

This chapter introduces Vygotsky's short life in its social and historical context and includes a section on those scholars who influenced his work. The more I learned about Vygotsky's life the more I improved my understanding of his teaching and research, which led me to adapt my own work to incorporate Vygotskian approaches in my learning, teaching and research in science education in the twenty-first century. Most of the chapter summarises my intensive reading on Vygotsky, and my discussions with Vygotsky's family members, Russian scholars who were taught by some of Vygotsky's students, and international colleagues with whom I have worked for many years.

The chapter concludes with an introduction to some twenty-first-century scientific research that demonstrates the importance of social constructivism in academic learning and indicates how Vygotsky's work has contributed to the basis for current science learning and teaching.

## 2.2 Vygotsky in Context

Vygotsky was born in the city of Orsha, Byelorussia (now Belarus) on November 5th, 1896. His family were Jewish, and Orsha was in a designated area, the Pale of Settlement, within which Russian Jews were required to live and work. The Pale



C. Murphy, *Vygotsky and Science Education*,
https://doi.org/10.1007/978-3-031-05244-6_2

was situated in the western region of Russia, in an area covering approximately one million square kilometres between the Baltic and the Black Sea. Vygotsky's short life was heavily influenced by the prevailing conditions in Russia (including political, economic, social and famine), which are highlighted throughout the chapter.

### 2.2.1 *Childhood and Schooling (1986–1917)*

Shortly after his birth, Vygotsky's family moved southwards to Gomel, a bigger city than Orsha within the Pale. His father accepted a post as department chief of the United Bank of Gomel. His mother was a teacher, who remained at home after the birth of their eight children. She taught her children languages, poetry and philosophy and often took them to the theatre.

Vygotsky was the second child and the oldest son. His intellect was prodigious from an early stage. He was educated at home by a private tutor, Solomon Ashpiz, from the age of 11 to 15. Ashpiz was a mathematician, also Jewish, and had spent time in Siberia for revolutionary activities when he was younger. He was known as a very kind and gentle person yet would only teach what would now be called 'gifted' children. Ashpiz' teaching method used a technique based on Socratic dialogue. He would ask a question and listen with his eyes closed as Vygotsky answered. Ashpiz would then open his eyes and ask the young Vygotsky to repeat certain sections of the answer until he got them correct. This method was successful: by late adolescence Vygotsky had mastered the fundamentals of a classical education, several languages, and Jewish history and culture.

By the age of 15, Vygotsky earned the nickname 'little professor'. He went to a private Jewish boys' secondary school when he was 15 and excelled in mathematics, history, philosophy, literature and drama. He was a key member, and subsequently the leader of a 'history circle' – a group of interested young friends and siblings who were trying to develop a philosophy of history. Vygotsky was enthusiastic about Hegel's views on history, particularly in terms of the dialectical formula of thesis, antithesis and synthesis. The 'history circle' applied this analysis to historical events. And so began for him the importance of history in development which later manifested itself in his CHT of the development of higher psychological functions (HPFs).

At the end of his schooling, Vygotsky achieved gold-level standard, which entitled him to attend university. Unfortunately, that year only 3% of Jewish applicants were allowed university places. Furthermore, instead of the best students (of which Vygotsky would have been one) being selected, a ballot was held to choose successful applicants. Vygotsky's friend Semion Dobkin writes of Vygotsky's upset when he heard this news, assuming that his name would not be drawn from the hat. Dobkin set a wager that if Vygotsky got into university, he would have to buy a volume of Bunin's poetry for him. Dobkin won: Vygotsky attended Moscow University and Dobkin got his book (Levitin, 1982).

Vygotsky enrolled in medicine for his parents' sake, but soon after changed to study law. In 1913–1917, he pursued full-time studies at Moscow State University in the Law Department. However, his passion was to learn more about philology, psychology, languages, and literature. So, in addition to Moscow University, he audited classes in the social sciences and humanities in the Moscow City People's University: a private, liberal, 'progressive' institution that accepted students from a diverse section of society (Yasnitsky & van der Veer, 2015). It was from this second university that Vygotsky published a philological treatise on Shakespeare's *Hamlet,* which he completed in 1916 at the age of 20. This work received international acclaim. He was a keen theatregoer and became a sharp theatre critic throughout his short life.

In 1917, Vygotsky graduated from *both* universities, in law from Moscow, and a Liberal Arts degree (literature, psychology, philology and philosophy) from Shaniavskii University. One of the key outcomes of Vygotsky's university learning was:

> *...admiration for the beauty of the Word – including the problems of understanding art and masterpieces of world literature, the complexities of language in its historical development, the intricacies of speech production and their interplay with thinking, emotions, personality, and culture – remained at the center of his interests throughout his life.* (Yasnitsky, 2011, pp. 111–112)

Vygotsky continued to publish theatrical and book reviews and maintained a very strong interest in the history of Russian Jews, in addition to his work on education and psychology. He always worked with several colleagues. As is the case with most famous scientists, he would not have developed his ideas in theory and practice without collaboration with others. From his early work within the 'history circle', he continued to work closely with peers throughout his life.

Yasnitsky (2011) identifies two clear lines from Vygotsky's early life which most strongly influenced his work. The first is Vygotsky's affection for 'the Word', which can be thought of as speech and language, developed through his voracious reading in childhood and his early studies in the humanities and social sciences. The second line was Vygotsky's concern with social injustice, partly because of his experience as a member of a Jewish family living in the Pale of Settlement. In addition, in post-Revolutionary Russia, Vygotsky was active, and enthusiastic in embracing the "call for creating a 'New Man' capable of overthrowing the social constraints of the capitalist 'Old World' of violence, inequity, and oppression, equally capable of overcoming the limits of his own biological nature" (Yasnitsky, 2011, p. 112).

### 2.2.2 Russia During the Years 1896–1917

During the early years of Vygotsky's life, there were several Russian pogroms against Jews, perhaps the most extensive being those between the years of 1903 and 1906. The Russian word 'pogrom' means to wreak havoc and demolish violently. The pogroms in Russia were manifest as anti-Jewish riots, causing more than 1000 deaths. Vygotsky's father tried to defend Jews during the pogroms, and in 1911, when Vygotsky was 15 years old, he (his father) was tried but acquitted for defending Gomel during a Czarist pogrom. In 1914, the First World War broke out, and in 1917, just as Vygotsky was finishing university and reached the age of 21, the Russian Revolution and Civil War had begun. Russia was beset with famine, plague and severe poverty at this time. Vygotsky moved back to Gomel and stayed there until the age of 28, when he moved to Moscow.

### 2.2.3 *Early Career and Home Life: 1917–1924*

Vygotsky's Gomel period (1917–1924) was instrumental in his later career as an experimental and developmental psychologist. Between the years of 1917 and 1919, his life was beset with tragedy. There was famine and plague, and one of his younger brothers, Dodik, fell afoul of TB. Conditions in Gomel forced Vygotsky's mother and himself to take Dodik to the Crimea, where the climate and sanatorium provided better chances for his recovery. However, their journey was interrupted due to fighting between White Army and Red Army soldiers, and they had to stop at Kiev. Dodik's condition worsened there, and Vygotsky's mother Celia started to show symptoms of TB herself. As she too weakened, they returned home to Gomel, where, soon after, 12-year-old Dodik died. Within a year, Vygotsky's other brother, Sergei, died of typhus. Their mother suffered from depression after the death of both sons and her own ill health. In 1919, the next tragedy arrived as Vygotsky realised that he also had contracted TB. During this period Vygotsky passed through a temporary crisis that was possibly caused by these untimely deaths and the aggravation of his own illness.

Vygotsky took up work in Gomel as a private tutor initially, and then started teaching in several Gomel institutions. One of the outcomes of the October Revolution had resulted in emancipation for the Jews, so Vygotsky was now allowed to teach. Formerly, Jews were forbidden to teach or take up any government jobs. Later in 1919, Vygotsky's cousin, David Vygodsky, returned to Gomel. David was 3 years older than Lev Vygotsky, and as the pair shared literary and historical interests, they started a publishing house called Ages and Days, and a magazine entitled *Veresk*, but this project was unable to continue, as paper was soon marshalled by the State.

Between 1919 and 1924, Vygotsky taught what might now be referred to as 'foundation studies', comprising literacy, numeracy, etc., in several institutions, such as the Labour School, The Workers' Facility, the local Teachers' College and a vocational school for pressmen and metallurgists. He worked also with homeless children and those with 'defects', now referred to as 'special needs'. He also taught courses in logic and psychology at the Pedagogical Institute, courses in aesthetics and art history at the local conservatory, and a course in theater at the local studio. Vygotsky carried out empirical research in addition to his teaching, and wrote his early formal articles on psychology, the first of which were published in 1924 (Yasnitsky & Van der Veer, 2015).

During this period, Vygotsky met, and in 1924, married Rosa Noevna Smekova. Later they had two daughters: Gita, born in 1925, an educational psychologist who died in 2010, and Asya, born in 1930, a biophysicist who died in 1985. Both daughters outlived each of their parents.

### *2.2.4 Russia During the Years 1917–1924*

During the years between 1917 and 1919, the town of Gomel was beset with unemployment and starvation. Until 1918, Gomel and other Belarussian territories were occupied by German forces. In March 1918, the Treaty of Brest-Litovsk marked Belarus' exit from the First World War and later established the Belarussian National Republic. In 1919, Local Bolsheviks established the Belarussian Soviet Socialist Republic (SSR) – the Red Army captured Minsk and pronounced it the capital of the Byelorussian SSR. Later, in 1922, the Belarusssian SSR became part of the then established Union of the Soviet Socialists Republics (USSR), led by Lenin (alias of Vladimir Ilyich Ulyanov). Lenin died soon after, in 1924, and was succeeded by Stalin.

### *2.2.5 Vygotsky's Last 10 Years 1924–1934*

In January 1924, Vygotsky presented a paper to the Second Psychoneurological Congress in Petrograd (now St Petersburg), on the relationship between conditioned reflexes and conscious behaviour (i.e., why psychology cannot ignore the fact of consciousness). The paper challenged Pavlovian concepts and raised the importance of conscious planning prior to action as being unique to humans. Vygotsky spoke fluently and confidently, and his talk impressed many in the audience, including one of his future close collaborators, Alexander Romanovich Luria. Subsequently, Vygotsky was invited to take up a position in the Moscow Kornilov Institute of Psychology, where he stayed until 1929, working mostly with Luria and Aleksei Nikolaevich Leontiev. In 1925, he published *Consciousness as a Problem of the Psychology of Human Behaviour* and presented at conferences all over Europe, including London, in July 1925. A year later, his book *Educational Psychology* was published. on his return to the Soviet Union, Vygotsky was hospitalised due to a relapse of TB and, having miraculously survived, remained an invalid and out of a job until the end of 1926. His dissertation on the Psychology of Art was accepted as the prerequisite of scholarly degree, which was awarded to him in the autumn of 1925, *in absentia*.

Vygotsky spent most of 1926 reading through the Russian and Western psychology literature, from which he framed his ideas and methodology in terms of addressing a 'crisis' in traditional psychology in its strictly biological approach, which considered all human behaviours as reflexes. He maintained that the traditional approach to psychology lacked a unified theoretical basis, a sound methodology, and a strong connection between theory and practice (Zavershneva, 2014).

Between 1926 and 1930, Vygotsky collaborated with distinguished scholars such as Luria and Leontiev to develop a new approach to the study of psychology, which was instrumental in trying to understand ways in which humans use objects as mediating aids in memory and reasoning, focusing on how children develop higher cognitive function, and studying how social and cultural patterns of interaction shape the forms of mediation and developmental trajectories.

In the early 1930s, Vygotsky began working towards establishing a psychological theory of consciousness, which remained unfinished when he died in 1934. Towards the end of this period, he experienced his own 'crisis' by identifying some flaws in his theorising. However, at this time, a generous offer from the Ukraine Government, resulted in members of his 'Vygotsky Circle', including his close collaborators Luria and Leontiev, moving from Moscow to Kharkov where they established the Kharkov School of Psychology. Another barrage of external criticism of Vygotsky came from his involvement in the holistic study of the child, known as 'paedology', an area considered by the Russian Government as 'bourgeois' and involving too much Western research, such as Gestalt psychology, and the work of Kurt Lewin and Kurt Goldstein.

One of Vygotsky's key contributions to research and practice was the interrelationship between language development and thought, which was published in his most famous work (*Thinking and Speech*, or *Myshlenie i Rech* in Russian) in 1934, shortly after his death. This book established the explicit and extremely strong connection between speech (both silent inner speech and oral language) and the development of concepts and cognitive awareness.

### 2.2.6 Vygotsky's Death and Its Aftermath

Vygotsky died of TB on June 11, 1934, at the age of 37, in Moscow. One of the last entries in his notebooks (collated and edited by Zavershneva and Van der Veer [2018, p. 497]) is:

> *This is the final thing I have done in psychology – and I will like Moses die at the summit, having glimpsed the promised land but without setting foot on it. Farewell, dear creations.* ***'The rest is silence.'*** (Emphasis in original, quoted by Vygotsky from *Hamlet*)

Immediately after his death, Vygotsky was proclaimed as one of the leading psychologists in the Soviet Union, and his brain was preserved for analysis of 'genius', along with the brains of others such as Lenin and Pavlov. However, the decree of the Central Committee of the Communist Party of 1936 denounced the mass movement, discipline, and related social practice of the so-called paedology, which resulted in some criticisms and *censorship* of Vygotsky's works. After the death of

Stalin in 1953, Vygotsky's works gradually came back into print, reaching the global publication of an early version of *Thinking and Speech*, entitled *Thought and Language*, translated by Hanfmann and Vakar, in 1962. This version omitted a lot of repetition from the original version in Russian and considerably reduced any reference to Marxism in the text. A further version edited by Kozulin in 1986, which was also flawed in ways such as reproducing elements of the 1962 text, included many errors in translation, as suggested in a review by Van der Veer (1987). The best version so far is *Vygotsky: Thinking and Speech*, translated by Norris Minick, in the first volume of the Collected Works of Vygotsky (1987).

Since then, several Vygotskian scholars have applied Vygotsky's experiments and ideas to education and psychology. Recently, Russian Vygotskian scholars have been studying the translations of Vygotsky's work and have identified issues of mistranslation and other editorial issues (e.g., see the works of Nikolai Veresov and Anton Yasnitsky). Today, Vygotsky's ideas have been widely appropriated by science educators' research and practice. Probably the most influential Vygotskian constructs which are used most extensively, include the ZPD, scientific concept development and CHT.

## 2.3 Key Influences on Vygotsky's Thinking

Apart from his colleagues, chiefly Alexander Luria and Alexeia Leontiev, and other collaborators, Vygotsky was influenced strongly by a wide range of literary authors, poets, playwrights, philosophers and psychologists. I have selected some of these to represent the variety and depth of influence drawn on by Vygotsky during his short life (Table 2.1). It is evident that much of his work was influenced by European and American scholars, which might have been considered anti-Soviet by the Russian Government in the Stalin era.

### *2.3.1 Psychology/Science/Philosophy*

Some of the strongest influences on Vygotsky's work on psychology, philosophy and education came from scholars who died before he was born, or when Vygotsky was too young to have been in contact with them, including Spinoza, Hegel, Marx, and William James. Vygotsky's philosophical stance could be described as 'high rationalism', influenced heavily by his readings of the Ancient Greeks, as well as Descartes, Leibnitz, Kant, and more especially the works of Spinoza, Hegel and Marx (Bakhurst, 2007). Such influence was manifest chiefly in Vygotsky's theoretical contributions to the fields of psychology, education, and other social sciences. It contributed to those aspects of Vygotsky's work on the swing from a purely biologically based understanding of human behaviour to the social-cultural explanation of human activity.

**Table 2.1** Some key influences on Vygotsky's research

| PSYCHOLOGY/SCIENCE/PHILOSOPHY | | |
|---|---|---|
| **Baruch Spinoza** | **1632-1677** | **Philosopher (HOLLAND)**<br>Baruch Spinoza was a Dutch philosopher of Portuguese Sephardi origin. He was one of the early thinkers of the Enlightenment and modern biblical criticism, including modern conceptions of the self and the universe. He was considered as one of the great rationalists of 17$^{th}$-century philosophy. |
| **Georg W. F. Hegel** | **1770-1859** | **Philosopher (GERMANY)**<br>Georg Wilhelm Friedrich Hegel was a German philosopher. He is considered the one of the most important figures in German idealism. |
| **Karl Marx** | **1818-1883** | **Philosopher, etc. (GERMANY)**<br>Karl Heinrich Marx was a German polymath: philosopher, economist, historian, sociologist, political theorist, journalist and socialist revolutionary. He was born in Trier, Germany, and studied law and philosophy at university. |
| **William James** | **1842-1910** | **Psychologist (USA)**<br>William James was an American philosopher, historian and psychologist. He was the first educator to offer a psychology course in the United States. |
| **Sigmund Freud** | **1856-1939** | **Psychoanalyst (AUSTRIA)**<br>Sigmund Freud was an Austrian neurologist and the founder of psychanalysis (a clinical method for treating psychopathology through dialogue between a patient and a psychoanalyst.) |
| **Pierre Janet** | **1859-1947** | **Psychologist (FRANCE)**<br>Pierre Marie Felix Janet was a pioneering French psychologist, physician, philosopher, and psychotherapist in the field of dissociation and traumatic memory. He is ranked alongside William James and Wilhelm Wundt as one of the founding fathers of psychology. |
| **Pavel Blonsky** | **1864-1941** | **Psychologist**<br>Pavel Petrovich Blonsky was a Russian Soviet psychologist and philosopher. He lived in Ukraine until 1918. He was one of the key theorists of Soviet paedology and introduced the behaviourist approach in Russian psychology. |

(continued)

**Table 2.1** (continued)

| | | |
|---|---|---|
| **Kurt Koffka** | **1886-1941** | **Psychologist (GERMANY)** Kurt Koffka was a German psychologist and professor. His major work was extending Gestalt theory (realisation of the whole as opposed to its parts). Gestalt psychology suggests that people experience life and learning as whole units and not as individual segments. |
| **Lev Vygotsky** | **1896-1934** | **Psychologist, educationalist (RUSSIA)** |
| **Jean Piaget** | **1896-1980** | **Psychologist (SWITZERLAND)** Jean Piaget was a Swiss psychologist known for his work on child development. His theory of cognitive development and epistemological view are together called 'genetic epistemology'. Piaget placed great importance on the education of children. |
| | | |
| **LITERATURE/DRAMA** | | |
| **William Shakespeare** | **1564-1616** | **Playwright (ENGLAND)** William Shakespeare was an English playwright, poet, and actor, widely regarded as the greatest writer in the English language and the world's greatest dramatist. |
| **Alexander Pushkin** | **1799-1837** | **Poet, playwright, novelist (RUSSIA)** Alexander Sergeyevich Pushkin was a Russian novelist, poet and playwright of the Romantic era. He is considered as the greatest Russian poet, and the founder of modern Russian literature. He was born into Russian nobility in Moscow. He died young (aged 37) in a duel. |
| **Fyodor Dostoyevsky** | **1821-1881** | **Novelist, philosopher (RUSSIA)** Fyodor Mikhaylovich Dostoyevsky (also spelled Dostoevsky) was a Russian novelist and short storywriter. His psychological penetration into the darkest recesses of the human heart, together with his unsurpassed moments of illumination, had an immense influence on 20th-century fiction. |
| **Leo Tolstoy** | **1828-1910** | **Novelist, educator (RUSSIA)** Count Lev Nikolayevich Tolstoy was a Russian writer who is regarded as one of the greatest authors of all time. He specialised in unconscious processes and described conscious mental life with unparalleled mastery. Tolstoy's name has become synonymous with an appreciation of contingency and of the value of everyday activity. |

(continued)

**Table 2.1** (continued)

| | | |
|---|---|---|
| **Anton Chekhov** | **1860-1904** | **Playwright, short story writer (RUSSIA)** Anton Pavlovich Chekhov was a Russian playwright and short-story writer who is considered one of the major literary figures of his time. Chekhov's plays take a tragicomic view of the staleness of provincial life and the passing of the Russian gentry. They have received international acclaim and are still staged world-wide. His short-story writing is unmatched. His overall body of work influenced important writers of many genres, including James Joyce, Ernest Hemmingway, Tennessee Williams, Henry Miller (also Lev Vygotsky). |
| **Konstantin Stanislavski** | **1863-1938** | **Theatre practitioner, actor (RUSSIA)** Konstantin Sergeyevich Stanislavski was a seminal Soviet and Russian theatre practitioner. He was recognised as an outstanding character actor. Also, the many productions that he directed garnered him a reputation as one of the leading theatre directors of his generation. |
| **Ivan Bunin** | **1870-1953** | **Poet, novelist, Nobel Laureate (RUSSIA)** Ivan Alekseyevich Bunin was the first Russian writer awarded the Nobel Prize for Literature. He was noted for the strict artistry which he carried on the classical Russian traditions in the writing of prose and poetry. |
| **Osip Mandelstam** | **1891-1938** | **Poet and writer (RUSSIA)** Osip Emilyvevich Mandelshtam (also spelled Mandelstam) was a major Russian and Soviet poet, prose-writer, and literary essayist. He was arrested by Joseph Stalin's Government during the repression of the 1930s and sent into exile with his wife. His works were not published during the Stalin era (1929-1953), and not released to Russians until the 1960s. Mandelstam and other associates founded the Acmeist School of Poetry, in an attempt to codify the poetic practice of the new generation of St Petersburg poets. Mandelstam summed up his poetic credo in his manifesto, *Utro Akmeizma* (written 1913, published 1919, *The Morning of Acmeism*). |
| **Lev Vygotsky** | **1896-1934** | **Theatre critic, art psychologist (RUSSIA)** |

In contemporary science education, the traditional consideration of scientific concepts as universal, permanent, and representative of 'truth', has moved towards the sociocultural view that scientific concepts are created in the scientific endeavour as tools to aid explanation, are subject to change, and are culture-dependent and context-bound. Science students are now encouraged to *use* scientific concepts in problem-solving activities, as opposed to rote learning and recalling such concepts.

Marx's conception of the essence of humans as "the ensemble of social relations" (Bakhurst, 2007, p. 58) provided the basis for Vygotsky's CHT. Vygotsky's methodological concept of a *unit of analysis* (*UoA*), referring to the smallest part of a whole which represents faithfully that whole, is derived from Marx. For example, the UoA for water *cannot* be represented truly by atoms of oxygen and hydrogen – which differ completely from each other and from water – but only by *a molecule of water*.

Vygotsky utilised Marxist concepts as foundational ideas for his general CHT and his empirical research; even his ZPD is arguably Marxist, in that it could represent the gap between the learner's present and the next stage towards becoming the 'New Soviet Man/Woman'. Vygotsky used methods and principles of dialectical materialism. He adopted Marxian-Hegelian dialectical thinking that genuine, ideal forms of life/behaviour need to be developed through social transformation. Vygotsky was also influenced by the philosophical considerations of Marx in his examination of the origin of human thinking and its social constitution by mediation and symbolic processes.

### *2.3.2 Consciousness*

William James' speculations on consciousness suggested that psychology as a field of empirical scientific research was, in Vygotsky's time, drifting towards a theory of consciousness that related it to language and meaning. Vygotsky's analysis of consciousness, particularly in the consideration of 'shared' consciousness, led him to consider word meaning as a UoA, such that if two people understood different meanings of the same concept, then they could not reach a shared consciousness of that concept. This is frequently the case in learning science contexts – a learner can recite the words attributed to a concept, for example, recite a definition of an atom, but not appreciate the functions of atoms in various chemical reactions. A comprehensive study of Vygotsky's changing ideas on consciousness can be found in Zavershneva (2014). To this day, there is not yet a satisfactory theory of consciousness, but Vygotsky's contribution to modern studies on consciousness is acknowledged, for example, by Koczanowicz (2011).

### *2.3.3 Development of Higher Psychological Functions (HPFs): From Social to Individual*

The work of psychologist Pierre Janet influenced Vygotsky's notion that sociocultural processes taking place in society and psychological processes taking place in the individual were connected. As explained by Van der Veer and Valsiner (1988, p. 58), Janet argued that all higher, typically human, behaviours have a social origin.

> *They exist first between people, as social acts, and only afterwards as intra-individual, private acts. These private acts, however, retain their social character.*

Vygotsky was also influenced by the *Gestalt* view, attributed to Kurt Koffka. *Gestalt is* a German word for form or shape, which may refer to Holism, the idea that natural systems and their properties should be viewed as wholes, rather than loose collections of parts. Vygotsky developed an approach that connects social and psychological processes and described the essential mechanisms of the socialisation and development of the human being (Gindis, 1999). This process was dependent on the cultural-historical context in which the development took place. Indeed, Vygotsky's general law of the development of HPFs states that:

> *...any function in the child's cultural development appears on stage twice, that is, on two planes. It firstly appears on the social plane and then on a psychological plane. Firstly, among people as an inter-psychological category [kategoria, in Russian] and then within the child as an intra-psychological category. This is equally true regarding voluntary attention, logical memory, the formation of concepts and the development of volition.* (Vygotsky, 1983, p. 145)

Vygotsky developed a mechanism for *how* higher-order learning created between people is appropriated by individuals. His idea was termed 'kategoria', a term used chiefly in Russian theatre and film, meaning a *dramatic collision*, which describes an inner tension causing a change in interest, motive or emotion and leads to change in behaviour (see Veresov, 2004). For Vygotsky, a dramatic collision must be experienced for the development of higher-order thinking, such as reflection. He argued that all that is taught is not always learned and does not necessarily lead to the development of HPFs, such as voluntary attention, reflection, and metacognition.

Pavel Blonsky, one of Vygotsky's teachers, also influenced Vygotsky's ideas on studying development historically. Blonsky maintained that behaviour is only intelligible as the history of behaviour. Vygotsky called his own method 'genetic experimental methodology', in which he studied the process of learning/development, as opposed to the outcomes of learning. Vygotsky examined complex systems in the process of change, using dialectical logic to understand the interrelationships between the components of the system, by using interventions and observing the learners' responses, for example:

1. An experimenter observes and reports participants' initial efforts to solve a 'new' problem (intervention) using their existing means.
2. The experimenter then introduces auxiliary means through which problem can be solved. This mediated assistance is observed and analysed theoretically and methodologically in terms of the learning process.

3. Developmental changes taking place are recorded over several sessions during which learners appropriate new psychological tools, leading towards development of HPFs.

### 2.3.4 *Child Development: Vygotsky and Piaget*

Piaget was born in the same year as Vygotsky, but they rarely met, chiefly because of the turbulence in Russia, and the fact that Vygotsky died at 37, whilst Piaget lived until he was 84. Child development is a rich and diverse discipline; among the most prominent theoretical perspectives on child development are Piaget's Stage Theory of Development and Vygotsky's Theory of Cultural-Historical Psychology (sociocultural model of development). These theorists proposed a genetic model of development and argued that children progress from lower to higher levels of development (Halldén, 2008). Vygotsky and Piaget were both cognition theorists who argued that children learn through creating their own experiences and both emphasised the importance of language, play and social interaction in the development of children, albeit in two very different ways (Halldén, 2008). More detail of Piagetian and Vygotskian theories as applied to learning science during children's early years (about 3–8 years old) can be found in Chaps. 4 and 5.

### 2.3.5 *Literary Influence*

The literary influence on Vygotsky's work helped Vygotsky to realise and demonstrate the importance of both emotion and communication between teachers and students to achieving effective science learning.

Vygotsky used some short stories by Ivan Bunin to demonstrate physiological measures (respiration rates) from listeners, which indicated their emotional responses to the stories. Vygotsky read the story[ies] out loud to his classes of labourers, whilst his research students counted the labourers' breaths throughout. They found that at certain critical points, the increased breathing rates indicated very strong emotion. Vygotsky contended later that giving people access to the greatest products of the human soul motivated their learning hugely. These products include great art, the wonder of scientific achievements, and the power of music, story and nature.

Apart from Shakespeare, most of his literary influence came from Russia. However, Vygotsky wrote his master's thesis on *Hamlet* in 1916, and his PhD was entitled *The Psychology of Art*, submitted in 1925. He argued that the artistic effect was created by conflicts between form and content (Van der Veer, 2007). The artists (writers, in this case) play with the form to create artistic devices, which lead readers to contradictory expectations, tension and other emotions that can result in a discharge of energy, described by Van der Veer as *catharsis* (2007, p. 41).

The poets, for example Pushkin and Mandelstam, influenced Vygotsky's work on consciousness and communication. Novelists, particularly Tolstoy, Dostoyevsky and Chekhov influenced other Vygotskian ideas, particularly in relation to the *condensed speech* that is frequently used by people who share each other's lives in different contexts. For example, in Tolstoy's Anna Karenina (Part 4, Chapter 13) such speech is illustrated in a declaration of love between Kitty and Levin:

> *...And he wrote down three letters. Before he had finished writing, she was already reading under his hand, and she finished the sentence herself and wrote the answer 'Yes'. Everything has been said in their conversation: that she loved him and would tell her father and mother that he would call in the morning.* (Vygotsky, 1986, p. 238)

In science learning and teaching, condensed speech is more likely to be used by one, say the teacher, but with a complete lack of shared understanding between teacher and learner(s). Part of the problem is the use of signs and symbols in science lessons which, just like words and phrases, may sound or read the same to learners and teachers, but there is little shared meaning.

Another reason (apart from the lack of shared meaning) that can form a barrier to shared understanding in science classes between learners and teachers can be explained by their respective motives. Vygotsky argued that thought is not an abstract, purely cognitive process; it is powered by desires, needs, interests and feelings; essentially, intellect and affect cannot be separated. Therefore, the motive for learning is key in developing conceptual science learning. According to Vygotsky, there is no deep learning without emotion/motive.

## 2.4 Vygotsky's Legacy and Twenty-First-Century Science Education

Vygotsky's work is becoming more and more visible in science education research and practice since there has been a significant shift from individual towards social learning in science. The top three science education research topics published in key science journals are (i) context of students' learning, (ii) science teaching, and (iii) students' conceptual learning since 1998 (Lin et al., 2019). Vygotsky's legacy to contemporary science learning is highly relevant in all these research and practice areas. For example, he highlighted the importance of the *cultural-historical context* (social and environmental) within which the science learning is taking place.

He demonstrated the necessity of spoken and written language to attain shared understanding between learners and teachers (*intersubjectivity*). Hence, collaboration between peers in the classroom, the importance of dialogue, and inquiry-based science learning are developments that are designed to promote shared understanding of science. Vygotsky also showed that learning leads development, which enables teachers to use the ZPD concept to help students move forward in their learning via inspiration, motivation, provocation and effective support.

(i) ***Context of Students' Science Learning***

Contemporary science classrooms differ from those in the late nineteenth and much of the twentieth centuries. Formerly, teaching was transmissive and much of the assessment was summative and based on recall of facts and procedures. Ideas from the Enlightenment period (from the late seventeenth to the early nineteenth centuries) suggested that human progress continued in a near-linear fashion and would lead to perfection. Unfortunately, that progress has instead resulted in damage to the planet Earth, so much so that without significant changes in human behaviour, escalating problems such as climate change are potentially catastrophic.

Science learning and teaching in formal contexts is now moving towards learning and teaching that will inform and upskill students to contribute to humanity's efforts to address these problems. We are developing a global outlook since many developments require collaboration between scientists to work together. We need to ensure that science and technological advances work together with social partnerships, in order that we can move towards attaining the wisdom needed to balance the knowing *how* with the knowing *why*.

Many science classrooms are now more collaborative, and inquiry based, adopting more social constructivist approaches. Students are likely to be more active in their learning. There is more dialogue between students, as well as student-to-teacher dialogue. The curriculum has also changed, with more emphasis on sustainability and conservation, and with the inclusion of science and technology studies (STS) – the study of how society, politics, and culture affect scientific research and technological innovation, and how these, in turn, affect society, politics and culture.

Sociocultural factors are highlighted in recent research in science education, in terms of how they impact on different ways of student learning. For example, studies on the positive impact of the use of gesture, music and informal science learning, and the potential negative effects of stereotyping, low motivation, and student anxiety.

Vygotsky's CHT has contributed a framework through which the changes in science learning and teaching can develop towards the social transformation that is required to stem the negative impact of various human activities. A United Nations (UN) report (2019) suggests that the key drivers include:

> *Increased population and per capita consumption; technological innovation which in some cases has lowered and in other cases increased the damage to nature; and, critically, issues of governance and accountability.*

CHT is described in the next chapter on Vygotskian key concepts, and their application in various formal and informal science learning and teaching contexts are included in Chaps. 4, 5, 6, 7, 8, 9, and 10.

(ii) ***Science Teaching***

Science teaching, CPD for in-service teachers and the development of new science teachers via ITE is also changing for the same reasons as described in the above paragraphs. Vygotsky's work focused on the idea of developmental process, and ways to progress through different stages or periods. His use of the ZPD concept suggest that development and learning are promoted when targeted at the progress the learner can make with support. It is in this zone that learning can be expedited – the learner is nearly there already. Vygotsky argued that development is never linear, but comprises regression, zigzags, gaps and sometimes conflict. Overall, he used the metaphor of the tide for development – it ebbs and flows but continues in the same direction, nonetheless.

Coteaching, where more than one teacher collaborates in the planning, practice and evaluation of lessons, has been demonstrated as an approach which enhances science teaching and learning. Murphy et al. (2015) based a coteaching model on Vygotsky's work, primarily in relation to CHT and the ZPD.

Other recent science teaching innovations include various combinations of flipped learning, inquiry-based approaches, active learning, social constructivist teaching approaches, games, informal science settings, problem-based and cooperative learning techniques, and a range of technology-supported innovations in the classroom. All of these can benefit from Vygotsky's legacy, particularly with regard to social learning and ZPD creation.

(iii) ***Students' Conceptual Learning***

Developing abstract scientific concepts is an area of science learning which can be enhanced using aspects of Vygotsky's legacy. Briefly, Vygotsky suggests that abstract concepts should be learned in terms of the dialectical relationship between concrete and abstract concepts. For example, the concrete concept of a pool of water (puddle) disappears via evaporation, an abstract concept. The two are inextricably linked, and the dialectical relationship is that learning about evaporation makes the concrete concept more abstract, whilst at the same time giving examples of evaporation, such as a puddle disappearing, makes the abstract concept more concrete! Teaching via this interplay leads to the improved learning of scientific concepts.

Vygotsky's ideas and research on concept development are treated in somewhat greater detail in the next two chapters, and again in Chap. 7, which has a focus on the learning of scientific concepts in post-primary school science.

## *2.4.1 Vygotsky and the Science of Learning*

Vygotsky's legacy to science education has also become more prominent in twenty-first-century research in the science of learning. For example, Meltzoff et al. (2009) carried out an interdisciplinary study on the science of learning by analysing the findings from the best research in four different fields: psychology, neuroscience, education and machine learning (also known as artificial intelligence).

Their key findings suggest that researchers in developmental psychology have, for many years, identified social factors that are essential for learning. Neuroscientists have identified brain systems involved in social interactions and mechanisms for synaptic plasticity that contribute to learning. Education research is continuously providing ideas and strategies which have been shown to enhance learning;

classrooms can be laboratories for discovering effective teaching practices. In machine-learning research, powerful learning algorithms have demonstrated that contingencies in the environment are a rich source of information about social cues (Meltzoff et al., 2009, p. 284). The research points to the requirement of social interaction for learning, suggested in the early twentieth century by Lev Vygotsky. Their overall conclusions can be summarised as:

> *A convergence of discoveries...leading to changes in educational theory...key component...role of 'the social' in learning.* ***What makes social interaction such a powerful catalyst for learning?*** *(Emphasis added) ... How can we capitalise on social factors to teach children better...? These are deep questions at the leading edge of the new science of learning.* (Meltzoff et al., 2009, p. 288)

Vygotsky's work has been used in many aspects of the science of learning, including neuroscience, digital learning, and the impact of the social and learning environments on the effectiveness of student learning. For example, An OECD publication: *Developing Minds in the Digital Age: Towards a Science of Learning for 21st Century Education* (Kuhl et al., 2019) drew together a corpus of innovative research, aimed at introducing the science of learning to those involved in education policy and practice. Vygotsky's ideas are included; for example, cultural tools for learning are now found on digital devices, and adult-child interactions provide engagements which support learning (Barron & Levinson, 2019). Vygotsky's ZPD is also adopted in the context of learning, which suggests that it is best to learn new skills when the learner is challenged, but not faced with very difficult tasks (Toub et al., 2019). Vygotsky's work is also considered in collaborative learning, which suggests that learners need to be engaged in socially embedded, and interest-driven, experiences; these experiences acknowledge and build on their past learning experiences (Suarez et al., 2019).

## 2.5 Summary and Conclusion

The first part of this chapter has summarised Vygotsky's life and legacy, illustrating his social and cultural context in terms of his short life and those who influenced his work. It has illustrated how and why Vygotsky was never able to progress his theory as far as he would have liked, due to the conditions in Russia at the time, which led to his early death and the subsequent denouncement of some of his work until the death of Stalin in 1953. His work was influenced by a wide variety of scholars in several fields, including psychology, philosophy, science, history, politics, literature and drama.

Vygotsky's legacy was then considered in terms of science learning and teaching in the twenty-first century, at a time when his work is gaining in popularity and influence. Three of the key aspects of Vygotsky's legacy to science education are: CHT, the ZDP and the nature and development of scientific concepts.

The next chapter further discusses the scholars who influenced Vygotsky – mostly in the development of his theoretical frameworks and methodology – which led to the development of the key Vygotskian principles and practices that we now employ in science education. Chapter 3 focuses on Vygotskian concepts and practices, and how they have been employed in science education.

## References

Bakhurst, D. (2007). Vygotsky's demons. In H. Daniels, M. Cole, & J. V. Wertsch (Eds.), *The Cambridge companion to Vygotsky* (pp. 50–76). Cambridge University Press.

Barron, B., & Levinson, A. (2019). Digital media as a catalyst for joint attention and learning. In P. K. Kuhl, S. S. Lim, S. Guerriero, & D. Van Damme (Eds.), *Developing minds in the digital age* (pp. 105–114). OECD Publishing. Retrieved from. https://doi.org/10.1787/c8b53b86-en

Derry, J. (2004). The unity of intellect and will: Vygotsky and Spinoza. *Educational Review, 56*(2), 113–120.

Gindis, B. (1999). Vygotsky's vision: Reshaping the practice of special education for the 21st century. *Remedial and Special Education, 20*(6), 333–340.

Halldén, K. (2008). The Swedish educational system and classifying education using the ISCED-97. *The international standard classification of education (Isced-97): An evaluation of content and criterion validity in 15.*

Koczanowicz, L. (2011). The dialogical concept of consciousness in LS Vygotsky and GH Mead and its relevance for contemporary discussions on consciousness. *Polish Psychological Bulletin, 42*(2), 65–70.

Kuhl, P. K., Lim, S. S., Guerriero, S., & Van Damme, D. (2019). *Developing minds in the digital age*. OECD Publishing.

Levitin, R. (1982). *One is not born a personality: Profiles of Soviet education psychologists.* Progress Publishers.

Lin, T. J., Lin, T. C., Potvin, P., & Tsai, C. C. (2019). Research trends in science education from 2013 to 2017: A systematic content analysis of publications in selected journals. *International Journal of Science Education, 41*(3), 367–387.

Meltzoff, A. N., Kuhl, P. K., Movellan, J., & Sejnowski, T. J. (2009). Foundations for a new science of learning. *Science, 325*, 284–288.

Murphy, C., Scantlebury, K., & Milne, C. (2015). Using Vygotsky's zone of proximal development to propose a better explanatory model for conceptualizing coteaching in preservice science teacher development. *Asia-Pacific Journal of Teacher Education, 44*(3), 281–295. Retrieved from. https://doi.org/10.1080/1359866X.2015.1060291

Suarez, N., Samano, M., Shuying, Y., Snyder, S., & Chukoskie, L. (2019). Research-practice partnerships in STEM informal environments. In P. K. Kuhl, S. S. Lim, S. Guerriero, & D. Van Damme (Eds.), *Developing minds in the digital age* (pp. 187–195). OECD Publishing. Retrieved from. https://doi.org/10.1787/c8b53b86-en

Toub, T., Verdine, B., Golinkoff, R., & Hirsh-Pasek. (2019). Shapes, blocks, puzzles and origami: From spatial play to STEM learning. In P. K. Kuhl, S. S. Lim, S. Guerriero, & D. Van Damme (Eds.), *Developing minds in the digital age* (pp. 177–186). OECD Publishing. Retrieved from. https://doi.org/10.1787/c8b53b86-en

Toulmin, S. (1981). Mozart in psychology. *Voprosy Filosofii, 10*, 127–137.

UN Report. (2019). *Nature's dangerous decline "unprecedented" extinction rates "accelerating"*. Retrieved from https://www.un.org/sustainabledevelopment/blog/2019/05/nature-decline-unprecedented-report/

Van der Veer, R. (1987). Thought and language. *The Journal of Mind and Behaviour, 8*(1), 175–178.

Van der Veer, R. (2007). *Lev Vygotsky*. Continuum International Publishing Group Ltd.

Van der Veer, R., & Valsiner, J. (1988). Lev Vygotsky and Pierre Janet: On the origin of the concept of sociogenesis. *Developmental Review, 8*(1), 52–65.
Veresov, N. (2004). Zone of proximal development (ZPD): The hidden dimension? In A. L. Ostern & R. H. Ylikallio (Eds.), *Language as culture tensions in time and space* (Vol. 1, pp. 13–30). ABO Akademi.
Vygotsky, L. S. (1983). *Collected works volume 3: Problems in the development in mind.* Izdatelistvo Pedagogika.
Vygotsky, L. S. (1934/1986). *Thought and language*. MIT.
Vygotsky, L. S. (1987). Thinking and speech. In *The collected works of L.S. Vygotsky, vol. 1* (pp. 37–285) (R. W. Rieber & A. S. Carton, Eds.; N. Minick, Trans.). Plenum Press.
Yasnitsky, A. (2011). Lev Vygotsky: Philologist and defectologist. A socio-intellectual biography. *Portraits of Pioneers in Developmental Psychology, 7*, 105–129.
Yasnitsky, A., & van der Veer, R. (2015). *Vygotsky: An intellectual biography*. Routledge.
Zavershnea, E., & Veer, R. V. D. (Eds.). (2018). *Vygotsky's notebooks: A selection* (Vol. 2). Springer.
Zavershneva, E. (2014). The problem of consciousness in Vygotsky's cultural-historical psychology. In *The Cambridge handbook of cultural-historical psychology* (pp. 63–97). Cambridge University Press.

# Chapter 3
# Key Vygotskian Concepts for Science Education

## 3.1 Introduction

This chapter introduces and explores aspects of Vygotskian theory in science education research and practice. Vygotsky began his educational research when he finished university, at the age of 20, when he became familiar with the work of other scientists. This chapter gives an overview of the key concepts and practices which arose from Vygotsky's research and teaching.

The chapter focuses chiefly on four key Vygotskian concepts that are most relevant to science education. These are: the ZPD, the nature and development of scientific concepts, CHT, and CHAT. In addition, the chapter will briefly introduce some of Vygotsky's lesser-known terms and concepts and how they relate to specific areas and contexts in science education. These concepts include: intersubjectivity, social situation of development (SSD), perezhivaniye, volition, and the unities of affect & intellect, real ideal, and learning and development. Certain concepts, such as social constructivism, have been attributed to Vygotsky, although he did not develop these himself. Some of his ideas have been utilised by researchers who promote social constructivism.

The terms: Vygotsky's ZPD, for example Bruner (1984), and Vygotsky's social constructivism', for example Schreiber and Valle (2013) have been interpreted wrongly, as they suggest that Vygotsky 'invented' the ZPD and social constructivism. The literature on Vygotsky and his work is contested by several scholars, who interpret Vygotsky's theoretical and practical in different ways. It could be argued that the complexity of studying Vygotsky and his work can lead to reductionist, simplified outcomes, such as 'tips for teachers'.

My own experience of studying Vygotsky and his work spans 30+ years, and I still struggle to clarify some of the theoretical arguments for concepts which have been attributed to him. Rather, I tend to put into practice ways of teaching and learning that I have observed in Russian Vygotskian schools, as well as experimenting with aspects of Vygtoskian theory.



C. Murphy, *Vygotsky and Science Education*,
https://doi.org/10.1007/978-3-031-05244-6_3

## 3.2 Zone of Proximal Development (ZPD)

Vygotsky best known for a very small fraction of his work, the notion of the ZPD. Indeed, Chaiklin (2003, p. 43) cites only eight published texts in which the phrase ZDP is used. The first mention is attributed to a lecture Vygotsky gave in March 1933 on an analysis of the pedagogical process (Vygotsky, 1933/1935, summarised in Van der Veer & Valsiner 1991, pp. 336ff).

### *3.2.1 What Is the Zone of Proximal Development (ZPD)?*

This is not the easiest question to answer, simply because, unlike Piaget (1896–1980) who lived long enough to explore and explain his ideas in full, Vygotsky (also born in 1896) died at the age of 37, relatively early in his research career. There is much discussion and debate about what Vygotsky meant by the ZPD. Indeed Palincsar (1998, p. 370) claimed that "… [the ZPD] is perhaps one of the most used and least understood constructs to appear in contemporary educational literature". She suggested that the original purpose of the ZPD, and its theoretical framework, have been misunderstood in the rush to use the ZPD as an explanatory tool, rather than to recognise its descriptive power in terms of the use of cultural tools and artefacts that are present in the learning activity to mediate learning in the ZPD. In this regard, Palincsar (1998) argues that people have taken too literally the idea of the more capable other in creating spaces for assisted performance. Veresov (2004) also suggested that, to understand the ZPD concept, it should be considered within the framework of Vygotsky's theoretical constructions.

The simplistic definition of the ZPD, seen in many textbooks and teaching guides, is of a 'gap' between what a child can achieve unaided, and with help. Such a definition could imply little more than the fact that teachers need to help children, seemingly stating the obvious. Mercer and Fisher (1992) point out the danger that the term ZPD is sometimes used as a fashionable alternative to the Piagetian terminology, which is more concerned with individual representations of learning, as opposed to learning as a social, culturally based process. Tudge and Scrimsher (2003) observed that in the six volumes of Vygotsky's collected works, the ZPD only appears on a few pages in the thousands he wrote. However, Van Oers (2007) discussed the complexity of the ZPD and showed how the concept was an *evolving* notion, even during the short research life of Vygotsky, used initially as an index for intellectual potential and later as an educational concept focusing on the conditions needed to *establish* a ZPD. Meira and Lerman (2001) also suggested that Vygotsky first used the ZPD as a means of testing intellectual development, but argued that the concept was later broadened by Vygotsky to examine the relationship between education and development (mainly through social assistance and play).

The ZPD has been variously described as a "*discrepancy* between a child's actual mental age and the level [she or] he reaches in solving problems with assistance"

(Vygotsky, 1986, p. 187). Vygotsky represented the ZPD as psychological "*functions* which have not yet matured but are in the process of maturing… 'buds' or 'flowers' of development rather than 'fruits' of development" (1978, p. 86), and a "*place* at which a child's … spontaneous concepts 'meet' the systematicity and logic of adult reasoning" (Kozulin, 1986, p. xxxv). Van Oers (2007, p. 15) pointed out that the ZPD "is *not* a specific quality of the child, *nor* is it a specific quality of the educational setting or educators… it is… collaboratively produced in the interaction between the child and more knowledgeable others". Wells (1999) and Tudge and Scrimsher (2003), together with many other researchers, also discussed the ZPD as an *interaction* between the students and co-participants. Meira and Lerman (2001) suggested that the ZPD is a *symbolic* space for interaction and communication where learning leads development. The interaction definition, whilst popular, has been contested. Chaiklin (2003) argues that the maturing functions described by Vygotsky (1978) are not created in an interaction, but that interaction helps to identify the existence of such functions and the extent to which they have developed.

Vygotsky contended that a full understanding of the ZPD should result in a reevaluation of the role of 'imitation' in learning. His notion of 'imitation' is not meant to be thought of as copying – more as emulation of an activity as part of the learning process. For example, children learning to add, knit or dance, emulate the teacher before doing the task by themselves. This type of activity coincides with the ZPD in the sense that it bridges what a child can do with help and then alone. In science lessons, learners can, for example, emulate the work of scientists, as they use fossil fragments to create a picture or model of an ancestral species.

Vygotsky developed the ZPD notion within CHT and described it as a 'zone' in which a complex array of interactions between people and with their environment. These interactions, if planned well, can lead to facilitating learning. Vygotsky introduced the ZPD concept to the fields of psychology and education.

Basically, the ZPD represents ways to enhance learning. For instance, when we are introduced to a new scientific concept which is abstract, it is often too difficult to utilise that concept in different contexts. It is at this stage of learning that external mediation, for example a teacher or another learner, or a demonstration is required. These interactions help learners to reach their potential within a particular level of development. In other words, the interactions should challenge the learner, but should be limited to their potential development. For example, if teaching a class of 13- to 14-year-old students about Newton's Laws of Motion, it is pointless to expect them to appreciate quantum physics, which is way beyond their understanding at this level. However, there might well be one or more students in that class who have studied physics in great depth, perhaps with guidance from friends, siblings and/or physics-loving adults, who will be able to understand quantum physics.

Vygotsky characterised the ZPD as "functions which have not yet matured but are in the process of maturing… 'buds' or 'flowers', rather than 'fruits' of development" (Vygotsky, 1978, p 86) and proposed that it represents "the domain of transitions that are accessible by the child" (Vygotsky, 1987, p 211). Many, researchers, though not all, have described the ZPD as interaction, which is collaboratively produced between learner and teacher.

### 3.2.2 How Is the ZPD Used?

Vygotsky suggested the ZPD is a "*discrepancy* between a child's actual psychological age and the level he (*sic*) reaches in solving problems with assistance" (Vygotsky, 1986, p. 187). We can take from this that learners get so far but need help to move further.

Vygotsky argued that an individual's actual level of development, as determined by independent performance such as an IQ test, does not cover the whole picture of learning development, but very frequently encompasses only an insignificant part of it. He suggested that responsiveness to mediation during the ZPD provides insight into the person's future development. Thus, what the individual can do one day with help, they can do tomorrow alone.

Vygotsky described the *creation* of ZPDs in terms of levels of assistance given to learners of the same 'actual' cognitive level (e.g., same IQ scores) in solving more difficult tasks. Despite being measured at the same level, one child could solve the task with very little help (the support is thus within their ZPD for the task), whilst another cannot solve it even after several different interventions designed to support the learning (the support is outside the learner's ZPD for this task). Such interventions which create ZPDs may include:

- Demonstrating the problem solution and see if the child can begin to solve it.
- Beginning to solve it and ask the child to complete it.
- Asking the child to solve the problem with the help of a child who is deemed more capable.
- Explaining the principle of the needed solution.
- Asking leading questions, analysing the problem with the child, etc. (Gredler & Claytor Sheilds, 2008).

Vygotsky considered performance on summative tests as an indication of the child's past knowledge and argued that "instruction must be orientated towards the future, not the past" (Vygotsky, 1962, p. 104). The following quote summarises Vygotsky's position on the ZPD:

> *Imagine that we have examined two children and have determined that the psychological age of both is seven years. This means that both children solve tasks accessible to seven-years old. However, when we attempt to push these children further in carrying out the tests, there turns out to be an essential difference between them. With the help of leading questions, examples, and demonstrations, one of them easily solves test items taken from two years above the child's level of [actual] development. The other solves test items that are only a half-year above, his or her level of [actual] development... This means that with the help of this method, we can take stock not only of today's completed process of development, not only the cycles that are already concluded and done, not only the processes of maturation that are completed; we can also take stock of processes that are now in the state of coming into being, that are only ripening, or only developing.* (Vygotsky, 1956, 446–447, as cited in Wertsch, 1985, p. 68)

Another important feature of the ZPD is the acknowledgement that learning is not a linear process. We experience all sorts of problems and interruptions which cause regression of progress. We then go back and try again (recursion). The graphic in Fig. 3.1 illustrates regression and recursion as key elements of the ZPD, in addition to guidance from more knowledgeable others (sometimes referred to as MKOs).

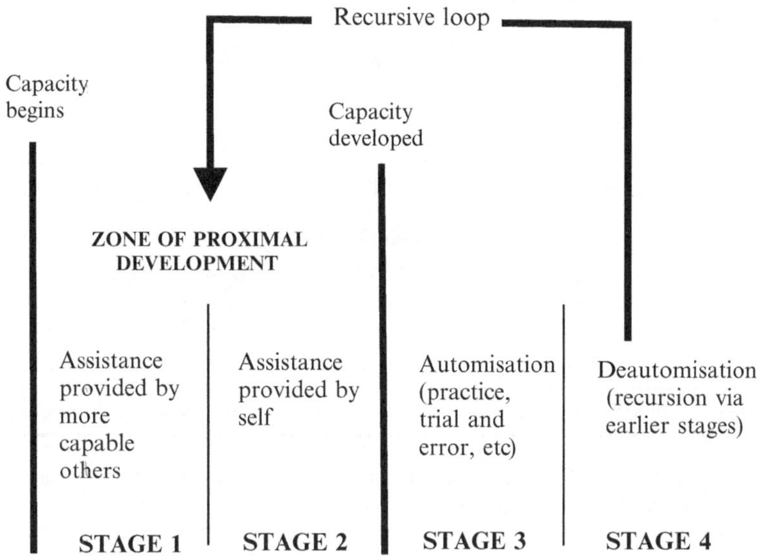


**Fig. 3.1** My adaptation of the ZPD by Tharp and Gallimore (1991)

Tharp and Gallimore's (1988) model of regression and recursion in the ZPD is useful in most contexts and will be referred to in many of the subsequent chapters in this book. Another important idea relating to regression and recursion is the use of Zebroski's (1994) metaphor of the tide, which moves both forwards and backwards, but the overall movement is forwards.

### *3.2.3 ZPD and Science Learning*

The literature concerning science learning before the twenty-first century is surprisingly neglectful of the work of Vygotsky; most emphasis is still placed on Piagetian ideas (Howe, 1996). Recently, however, there has been a growth in popularity of Vygotsky's ideas in general, which relates to preparing learners to contribute to a global, as opposed to a local economy. Students are required to develop skills relating to knowledge creation in collaborative networks, as opposed to learners in the past who needed to know facts and procedures. A Vygotskian approach to learning in a sociocultural framework is considered to better prepare the type of learner who will succeed in a knowledge-based economy, concerned with promoting a sustainable planet, among other challenges. The ZPD is a concept which, in school settings, is concerned with *moving the learner(s) on* to the next level. It can be used to enhance science learning by making school science more meaningful in terms of its situation within the science world and science in everyday life.

We can directly relate the ZDP as formulated by Vygotsky in relation to child development to the process of learning science. The 'zones' can be created to move the learner between school science, the world of science and the everyday world,

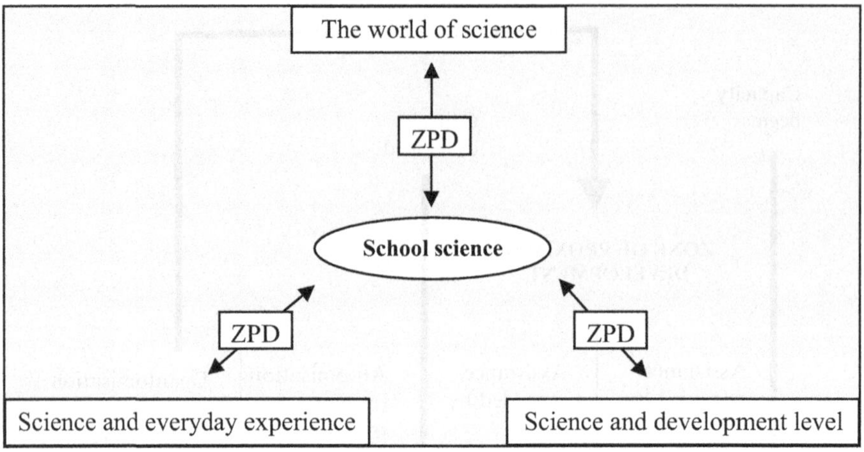


**Fig. 3.2** Sense frame in which school science can be made more meaningful. ZPDs can be created to link school science to the framework

and to relate most closely to their levels of development, that is, a 'sense-frame' for school science (see Fig. 3.2).

Within these zones, teachers, advisers, curriculum developers and policymakers can collaborate to develop frameworks, so that existing science curricula can be tailored, such that content and activities are characterised in terms of their explicit links to the world of science and the child's current stage of development, as well as links to everyday life.

Vygotsky described the creation of ZPDs in terms of levels of assistance given to learners of the same 'actual' cognitive level (e.g., same IQ scores) in solving more difficult tasks. Despite being measured at the same level, one child might solve the task with very little help (the support is thus within their ZPD for the task), whilst another may not solve it even after several different interventions designed to support the learning (the support is outside the learner's ZPD for this task). Such interventions may include: demonstration of the problem solution and see if the child can begin to solve it; beginning to solve it and ask the child to complete it; asking the child to solve the problem with the help of a child who is deemed more able; explaining the principle of the needed solution, asking leading questions, analysing the problem with the child, etc. (Gredler & Claytor Sheilds, 2008). Vygotsky considered performance on summative tests as an indication of the child's past knowledge and argued "instruction must be orientated towards the future, not the past" (Vygotsky, 1962, p. 104).

So how can the ZPD contribute to science learning in the twenty-first century? For optimal school science learning, teachers, administrators, curriculum developers, etc., might benefit from considering, in depth, the links between learning and development, and prepare content and pedagogy in the creation of ZPDs to support the learning. Kravtsov (2009, personal communication) indicated such links for science learning from early childhood until the end of second-level school education. Below is a summary of some of the key developmental stages described by Vygotsky and his followers and the links to science learning. Vygotsky rejected ideas of child development which focused on one specific indicator, such as teeth (Pavel Blonsky's stages of development were: no teeth, milk teeth, and then new teeth, which last through adulthood), and cognition as an indicator of developmental changes, proposed by Piaget. Vygotsky's concept of development was holistic, combining social, physical, cognitive, consciousness and other factors. Vygotsky argued that learning and development were fully interlinked. He argued that no single indicator, such as a level of cognition, can describe a child's ability to learn. Learning can be almost impossible for those whose learning environment is not inappropriate, such as a perceived bad teacher, boring delivery of content, difficult issues at home, etc.

A useful summary of the ZPD in science learning comes from Wass and Golding (2014, p. 671), which suggests that the basic idea is that "...we should pitch what we teach so that it is slightly too hard for students to do on their own, but simple enough for them to do with assistance". However, in different forms of development this idea could be contested on the grounds that it assumes a one-way relationship between 'learner' and 'teacher'. A social constructivist approach to science education research and practice positions learners and teachers *both* contributing to ZPD development. The ZPD concept is relevant to science learning at all levels and in many different contexts. The chapters in Part II of this book give examples of ZPD use in different levels and contexts.

### 3.2.4 Where Does the ZPD Begin?

A question arises about how and where the ZPD 'begins'. What *drives* development via the ZPD through one stage to the next? In many contexts, an answer can be found in the lesser-known Vygotskian concept of the SSD.

### 3.2.5 Social Situation of Development (SSD)

The SSD is a concept which is not used as frequently in the literature describing student learning in formal and non-formal contexts, although as a concept it could be very useful in terms of ZPD creation in science learning, preferably via both learner(s) and teacher(s).

The SSD can be conceptualised as the *tension* between what the learner wants to achieve and what is needed to achieve it. It has been described most frequently in the field of child development. An example could be an infant who can crawl, but not yet walk, becoming frustrated by the actions of a toddler who can reach a desired toy more quickly by running. This tension provides a starting point for the infant to mimic the movements of the toddler and eventually develop the capacity to walk and run. Once a new stage is reached, the child is in a new SSD characterised by the new psychological functions, and development starts again and reaches a crisis before progression to the subsequent stage.

At the start of each developmental stage, the child has a specific relation to the reality surrounding them (including their relation to carers and other people), which is characteristic of the child's age, and which reflects their level of development. The reality surrounding the child is culturally and historically determined, so is different for each child, depending on their historical period and specific cultural norms. This child-reality relation is the SSD and could be said to limit the child, who acts in ways to increase their agency in relation to their surroundings, driving forward progression to the next stage.

### *3.2.6 Perezhivanie*

Another Vygotskian concept which could be considered as a part of the ZPD is *perezhivanie,* which could be translated as a "conscious lived experience". It can refer to an unforgettable emotional experience of a child/person that has arisen from any situation or any aspect of the environment, which can impact on navigation through the ZPD. It describes the importance of emotion in learning. In the words of Vygotsky (1998, p. 291):

> *At the age of seven years, we are dealing with the onset of the appearance of a structure of perezhivanie, in which the child begins to understand what it means when he says, 'I'm happy,' 'I'm unhappy,' 'I'm angry,' 'I'm good,' 'I'm bad,' that is, he is developing an intellectual orientation to his own perezhivaniya (developing perezhivanie).*

*Perezhivanie* provides yet another argument for the importance of *relevance* in science learning, and how making science learning relevant, not just to a student's everyday life (such as the use of kitchen liquids as part of the teaching of pH), but also the importance of the pH concept in global issues such as climate change, and the use of the pH concept in many different areas of science research. Such meaningful learning promotes student agency, such that they have a more active, enriched experience of science learning.

The next section of this chapter introduces Vygotsky's ideas on scientific concept development. He proposes that such concepts are created by scientists, who have studied phenomena for a long time, until they can explain such phenomena in terms of other concepts. For example, the concept of evaporation can be explained in terms of energy, wind, time, and temperature. Concepts are changed when there is new instrumentation, knowledge and/or different contexts. Vygotsky maintained that learning definitions of concepts does not lead to understanding.

## 3.3 Vygotsky and Scientific Concept Development

Vygotsky's work on concept development is most important for science education research and practice. To start, what are scientific concepts? Most definitions refer to a scientific concept as an idea, or a law, which helps to explain a phenomenon under investigation.

Vygotsky proposed that human cognitive development is different from that of animals, in that it is largely, although not entirely, based on language. This collective memory, or knowledge, was externalised with the invention of writing. Knowledge was more permanent and thus independent of who produced it and became written 'objects' used for education in different cultures, and passed on via collaborative activity using cultural tools, such as pictures, diagrams and writing.

As science developed, concepts were created, based on empirical observation and thorough scientific investigation, to help explain phenomena. Cultural tools are the signs and symbols which comprise the mechanism for the development of higher cognitive skills, or, in today's parlance, *thinking skills* (Gredler & Claytor Shields, 2008). They include graphs, charts, symbol systems and language.

Vygotsky proposed a 'super-concept' framework of four major concepts, which define *all* human activity in the world (Kravtsova, 2010):

- *Time* – all human activity in the world occurs in a certain time.
- *Space* – all human activity takes place within a space, or place.
- *Substance* – all human activity uses substance, or materials.
- *Conscious reflection* – human activity differs from other animals because of the element of reflection on what, how and how to improve, the action or activity.

This framework provides a structure within which every scientific concept can be subsumed. The place for many of the 'process' concepts can be found within Vygotsky's framework under 'conscious reflection'.

Vygotsky linked concepts to each other using a metaphor that is based on the surface of a globe, onto which concepts can be placed using a system of coordinates, corresponding to latitude and longitude in geography. A concept's 'longitude' relates to its degree of abstraction, and thus characteristic of thought processes, whilst its 'latitude' represents its objective reference, for example, plant or animal (Vygotsky, 1986, pp. 199–200).

Vygotsky contended that in a *true scientific concept*, the bonds between the parts of an idea and between different ideas are logical; thus, the ideas form part of a socially constructed and accepted system of hierarchical knowledge (Berger, 2005). Within this system, concepts can be regarded as tools, which help scientists to explain the world.

Some other concept types which are important to science learning and practice are less concrete, for example *inference*, or investigation. These concepts enable a broader definition, which also leads to the idea of concepts as 'tools'.

Vygotsky's ideas form an *active* description of scientific concepts, in that they are constantly being tested for their ability to function as tools in different contexts. Some tools are better than others at doing a specific job. It could be the case that some scientific concepts serve the science context well, but not the science education context – for example, respiration. The term 'respiration' is frequently confused with 'breathing' by younger learners, and the biochemistry of respiration is far too difficult for many senior school biology students to understand, unless they have a very good knowledge of chemistry.

Some concepts are very tricky to use, especially if there is complex mathematics involved, such as relativity theory. Are the scientific concepts which can be used in schools for science learning fit for purpose? Or is the question: is the *way we teach* SCs fit for purpose? It can be useful for students to be made aware that each scientific concept has been generated during the investigation of specific contexts and then replicated to test its generalisability.

SCs are not permanent; however, they change with time and new technologies. Fensham (2015) argued for learners to become "connoisseurs of science" (p. 35). His ideal can be developed in school science with increased attention to discussions of how, where and when various scientific concepts came about, including the associated difficulties, political and technological barriers, and enablers, as well as other human factors, to engender a deeper appreciation of the scientific endeavour.

### *3.3.1 Developing Scientific Concepts*

The key Vygotskian idea of concept development is that the process is *not linear*, despite the popular notion of progression in concept formation as a form of instruction. Rather, Vygotsky (1986) proposed a dialectical model for the development of scientific concepts: "the child's scientific and [their] spontaneous [everyday] concepts… *develop in reverse directions* [original italics] … they move to meet each other" (p. 192). Using the example below in Fig. 3.3, the students' everyday concept of a *puddle* (pool of water) develops more scientifically when they learn about evaporation; at the same time, their concept of *evaporation* will become more 'everyday' to them when applied to familiar contexts, such as puddles and perspiration.

A major 'take home' message from concept formation is that most of the assessment in science is aimed at students who have reached the level of 'pseudoconcepts', as opposed to true conceptual understanding. Pseudoconcepts, can be confused with true concepts because the learner might be *using the right words to describe the concept*, but lacking the logical connections between its parts.

For example, the learner may *recite the definition of an ionic bond to describe how it differs from a covalent bond* without understanding the nature of chemical bonding. The learner is able to use pseudoconcepts in communication and activities, such as exams, as if they were true concepts. The words of the learner and teacher may refer to the same idea, but their meanings may not be the same (Gredler & Claytor Shields, 2008). Berger (2005) suggested that true concepts are formed from pseudoconcepts via the *appropriate* use of signs and social interventions (frequently the teacher), thereby creating a bridge between the individual and social meanings. A true concept is bound by *logical* bonds within parts and between different concepts.

Vygotsky and his co-workers studied the ways that learners of different ages struggled or successfully used these aids, documenting changes in learner activity and accompanying changes in cognitive functioning. The task was to sort a set of wooden blocks of different colours, shapes and sizes into four groups. The experiment is now called "The Vygotsky Blocks" experiment, which led to the proposal of the stages that are passed through in the formation of concepts by children (See Chap. 4, Sect. 4.2 (ii) for more details).

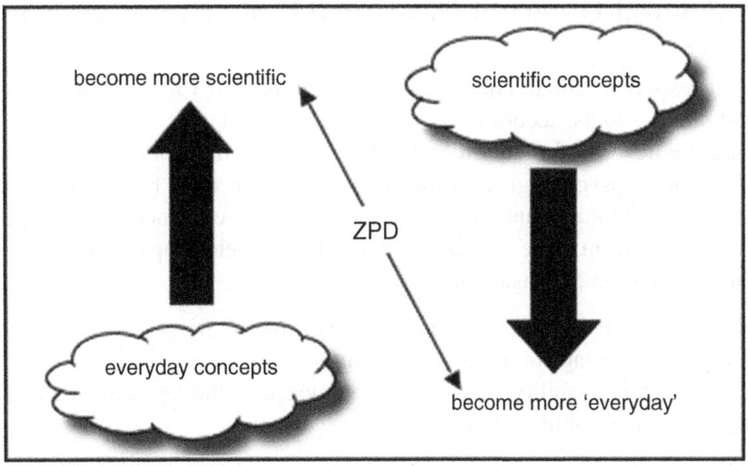


**Fig. 3.3** A dialectical model of concept development

The stages comprise:

1. Random grouping.
2. Thinking in complexes.
3. Development of pseudoconcepts.
4. Understanding and using scientific concepts.

A major problem in concept development is recognition of when true conceptual thinking is being demonstrated. Unless this process involves evidence that learners are *using* the concept(s) appropriately, it could be argued that it is a pseudoconcept, not a true concept.

Another issue is the case that learners can be thrust into problem-solving with new concepts before they have developed them sufficiently for the task, resulting in incomplete concept formation. The eventual formation of true concepts indicates that the learner is now able to master their own thinking. One of the most difficult tasks for learners, according to Gredler and Claytor Shields (2008) is to learn the connections and relationships between concepts. The advice is for teachers to construct a large visual diagram of the concepts in the topic, and between topics, as the term progresses. This activity requires pre-planning by the teacher to identify the required concepts for learning in advance.

Many examples of how science teaching can lead to learners' formation of true concepts are described in the chapters in Part II of this book. The move towards more collaborative and cooperative learning strategies in which students are encouraged and facilitated to repeat experiments as required, mimics more closely the science world that they may wish to enter. Essentially, if teachers try to move from teaching the curriculum towards teaching students by engaging their interests and relating that work to the curriculum, there is much more potential for significant improvement in their scientific concept development and motivation to learn.

## 3.4 Cultural-Historical Theory (CHT) in Science Learning

CHT suggests that human cultures develop *cultural tools* to solve problems. The greatest of these tools, according to Vygotsky, is language.

Vygotsky developed the term 'cultural tools' from his work with children who had physical or psychological problems. He saw that their biggest problem was isolation from the dominant cultural heritage. He was convinced that they needed different 'tools' from those used by most people to develop capacity to work well in the dominant culture, for example:

- Blind – braille (cultural tool) – we could ALL use it…
- Deaf – sign language (cultural tool) – ditto.
- Psychological 'disability' – variety of treatments, including medicines and therapies are used as cultural tools.

Formal science learning can be considered as a cultural tool. In science learning, we can recognise cultural tools such as schools, specialist teachers, books, digital and laboratory resources. In CHT, the tools *mediate* the learning, which serve to enable the learner to think, understand and to use scientific concepts to develop their awareness and ideas to conserve, and perhaps improve, the world around us. Such learning requires higher-level thinking than that of our everyday activities. For higher-level thinking, we need to develop HPFs. Vygotsky identified such HPFs, such as:

- Voluntary attention.
- Reflection.
- Memory.
- Abstraction.
- Judgement.
- Insight.
- Orientation.

These and other HPFs are developed as learners utilise mediation resources (cultural tools) provided by teachers as *psychological* tools, which with practice, help to develop conscious awareness of their own learning, that is, metacognition. It is the interaction of a group of HPFs that we can use to apply and test solutions to problems. Figure 3.4 represents this process, using reflection as a HPF.

CHT deems that we need to develop higher-level thinking, which can start at a very early age. For example, very young children exhibit abstract thinking when they use neutral materials in their play to represent something in their play, such a stick to 'ride' as a horse or perhaps a box, to use as a garage, hospital ward or a shop. Such abstraction (e.g., from a stick to a horse) prepares children to use symbolic representation in later science learning. This early use of symbols is key to developing abstract thinking in science learning.

In formal learning, Vygotsky developed a general law of the cultural development of HPFs, which states:

> *Every function in the cultural development of the child comes on to the stage twice – that is, on two planes. It firstly appears on the social plane and then on a psychological plane. Firstly, among people as an inter-psychological category and then within the child as an intra-psychological category…* (Veresov, 2004, p. 2)

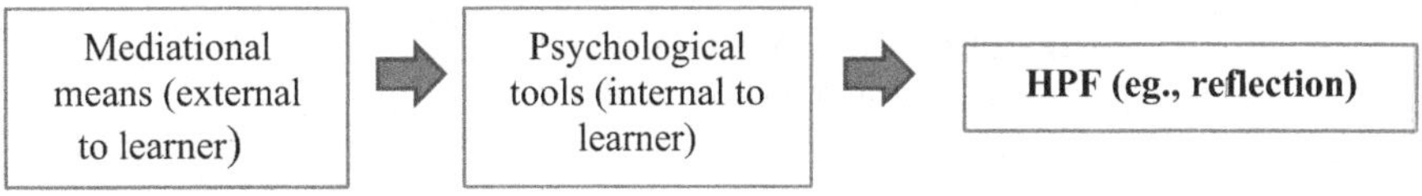


**Fig. 3.4** Development of a higher psychological function (HPF)

### *3.4.1 How Does This Learning Happen?*

Vygotsky frequently used theatrical terms in his teaching, combining his love of theatre and of his teaching and research with children and adults. A theatrical term he used in the above quote is *category*. Translated from early Russian (prior to 1917), the term 'category' was used as a dramatic collision, or a dramatic social situation. In the general law above, Vygotsky referred to category as a dramatic collision between the social (more than one person) and the learner as individual. The dramatic collision makes a learner think deeply, and sometimes change their behaviour. For example, a student in a science class might shout out a wrong answer, and a positive response from the teacher might help their learning by making them feel more motivated. A neutral teacher response might result in a student losing interest, whereas a negative response could have a negative effect on a student's motivation.

## 3.5 Cultural-Historical Activity Theory (Chat)

CHAT is a concept which arose from Vygotsky's observation that all actions are mediated between the subject and object of the action. The foundational concept of the CHAT framework is the *activity*, which is understood as purposeful, transformative, and developing interaction between actors (subjects) and the world (objects). CHAT was originally developed through a collaboration between Vygotsky and his colleague Leont'ev, based largely on Vygotsky's theories of cognition and learning.

Their so-called *first-generation* activity theory model centred on the premise that individuals do not merely react to their environment, rather their relationship to their environment is mediated by *tools and culture*. Vygotsky and Leont'ev offered an analysis of human activity which involved *subjects* and the *tools* they developed to work on *objects* of activity to achieve an *outcome* (see Fig. 3.4).

The example above in Fig. 3.5 illustrates the activity of learning and teaching. The *subject* could be one or more teachers. The *object* is the science lesson, or other ways of promoting science learning. The 'big picture' of the *outcome* of the activity could be that learners will become well science-educated citizens.

CHAT focuses on the *activity* of, say, teaching/learning, to provide an analytical framework which enables researchers to investigate the interactions between actors, tools, and practices, and how they interact during the process. The limitation of the first generation was that analysis remained largely focused on the key actors (e.g., students and teachers).

This limitation was overcome by the *second-generation model* in which Leont'ev expanded on the individual action in relation to the wider community. However,

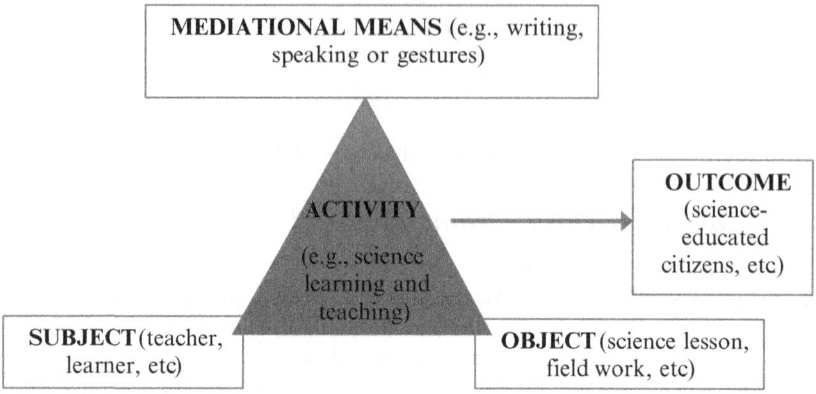


**Fig. 3.5** First-generation CHAT model

more recently 'Western' thinking in CHAT has been influenced mostly by the work of Yrjö Engeström, who further developed the second-generation model by expanding the original triangle (Fig. 3.4, above) to examine the social/collective elements of the activity. Engeström (1987) drew together the ideas of Vygotsky and Leont'ev and developed the second-generation activity theory model. He expanded the triangle to include the community in which activity occurs (see Fig. 3.5). Activity is collective, oriented towards an object and mediated by tools and signs with the main elements of activity being subject, object, mediating artefacts, community, rules, and division of labour.

CHAT enables the study of practice, as a group of people conducting a collective activity with a specific object or goal, rather than focusing on the individual, *per se* (Bødker, 1989).

Engeström (1987) produced a framework from which researchers can articulate the relationships between subject, tools, object, and outcome for activities such as science learning and teaching, and science education research. He further developed the idea of the 'activity system', as illustrated in Fig. 3.6. However, an activity system comprises the perspective and cultural characteristics of the s*ubject*. For example, the subject might be a teacher, or group of teachers, involved in the activity of implementing inquiry-based science education (IBSE) as a key area of their science pedagogies, with the outcome of developing students' critical thinking. If, instead of the teacher(s), the subject were school administrators, the outcome might be improving overall school performance. The two outcomes are not the same, so the interaction between these two subjects could lead to contradictions, which would potentially limit the outcomes for both. The *third-generation activity theory* model considers interactions between activity systems and is commonly used in researching inter-agency and other more complex cultural contexts. Third-generation activity theory models support the idea of networks of activity within which contradictions and struggles take place (Daniels, 2004).

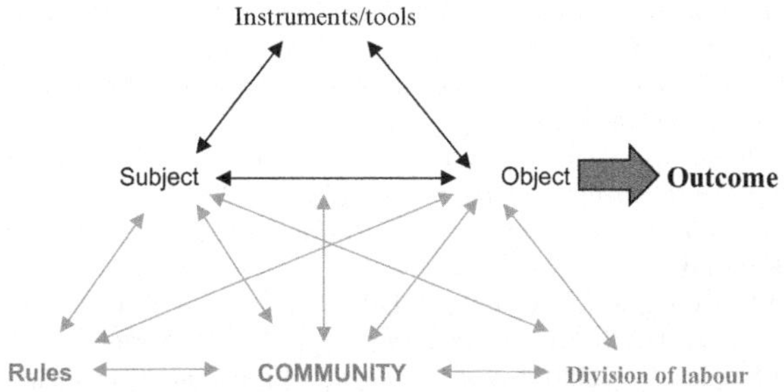


Key: The expanded CHAT triangle is illustrated in a different font colour, to highlight the **community** activity (e.g., school administrators and other teachers, parents). To work together, the subject(s) and community must follow 'rules' to ensure that things run well. The 'division of labour' fits in with the 'rules', such that the appropriate people work together most effectively.

**Fig. 3.6** Second-generation activity theory model (activity system), redrawn from Engeström (1987)

## 3.6 Summary and Conclusion

This chapter has provided a selection of Vygotskian ideas, concepts and constructs which offers a basis for their application in different science education contexts. Some will be familiar, such as the ZPD and non-linear concept development, but others less so. The lesser-used ideas are key in more specific contexts and can be applied further when more is known about them, in terms of current science education and research.

The next chapter applies the use of some of the Vygotskian concepts to science education research, both his own methodologies and those of science, psychology and education researchers. The high popularity of Vygotsky's work in science education research came about in the late twentieth century, along with the recognition of the need for increased collaboration in learning, teaching and research, to address the need for more global, as opposed to local, projects which can help society to deal with global problems such as climate change, environmental pollution, and non-equitable resource distribution.

## References

Berger, M. (2005). Vygotsky's theory of concept formation and mathematics education. In H. L. Chick & J. L. Vincent (Eds.), *Proceedings of the 29th conference of the international group for the psychology of mathematics education* (pp. 153–60). University of Melbourne.

Bødker, S. (1989). A human activity approach to user interfaces. *Human-Computer Interaction, 4*(3), 171–195.

Bruner, J. (1984). Vygotsky's zone of proximal development: The hidden agenda. *New Directions for Child and Adolescent Development, 84*(23), 93–97.

Chaiklin, S. (2003). The zone of proximal development in Vygotsky's analysis of learning and instruction. In A. Kozulin, B. Gindis, V. S. Ageyev & S. M. Miller (Eds.), *Vygotsky's educational theory in cultural context* (pp. 39–64). Cambridge University Press. Retrieved from https://doi.org/10.1017/CBO9780511840975

Daniels, H. (2004). Cultural historical activity theory and professional learning. *International Journal of Disability, Development and Education, 51*(2), 185–200.

Engeström, Y. (1987). *Learning by expanding: An activity-theoretical approach to developmental research*. Orienta-Konsultit.

Fensham, P. (2015). In D. Corrigan, C. Buntting, J. Dillon, A. Jones & R. Gunstone (Eds.), *The future in learning science: What's in it for the learner?* (pp. 35–59). Springer.

Gredler, M., & Claytor Sheilds, C. (2008). *Vygotsky's legacy: A foundation for research and practice*. Guildford.

Howe, A. (1996). Development on science concepts with a Vygotskian framework. *Science Education, 80*(1), 35–51.

Kravtsov, G. (2009). A cultural-historical basis of the "Golden Key" program. In *Paper presented at the International Psychological Vygotsky Summer School*. Belaya Kalitva.

Kravtsova, E. E. (2010). The cultural-historical foundations of the zone of proximal development. *Journal of Russian and East European Psychology, 47*(6), 9–24.

Meira, L., & Lerman, S. (2001). The zone of proximal development as a symbolic space. South Bank University, Faculty of Humanities and Social Science. In A. Kozulin, B. Gindis, V. S. Ageyev, & S. M. Miller (Eds.), *Vygotsky's educational theory in cultural context* (pp. 39–64). Cambridge University Press. Retrieved from https://doi.org/10.1017/CBO9780511840975

Mercer, N., & Fisher, E. (1992). How do teachers help children to learn? An analysis of teachers' interventions in computer-based activities. *Learning and Instruction, 2*(4), 339–355.

Palincsar, A. S. (1998). Social constructivist perspectives on teaching and learning. *Annual Review of Psychology, 49*, 345–375.

Schreiber, L. M., & Valle, B. E. (2013). Social constructivist teaching strategies in the small group classroom. *Small Group Research, 44*(4), 395–411.

Tharp, R. G., & Gallimore, R. (1991). *Rousing minds to life: Teaching, learning, and schooling in social context*. Cambridge University Press.

Tudge, J., & Scrimsher, S. (2003). Lev S. Vygotsky on education: A cultural-historical, interpersonal, and individual approach to development. In B. J. Zimmerman, & D. H. Schunk (Eds.), *Educational psychology: A century of contributions* (pp. 207–228). Lawrence Erlbaum Associates.

Van der Veer, R., & Valsiner, J. (1991). *Understanding Vygotsky: A quest for synthesis*. Basil Blackwell.

van Oers, B. (2007). In the zone. In *Children in Europe: Vygotsky edition* (pp. 14–15). Children in Scotland.

Veresov, N. (2004). Zone of proximal development (ZPD): The hidden dimension? In A. L. Ostern & R. H. Ylikallio (Eds.), *Language as culture tensions in time and space* (Vol. 1, pp. 13–30). ABO Akademi.

Vygotsky, L. S. (1962). *Thought and language*. MIT.

Vygotsky, L. S. (1978). *Mind in society: The development of higher psychological processes* (M. Cole, V. John- Steiner, S. Scribner, & E. Souberman, Eds.). Harvard University Press.

Vygotsky, L. S. (1986). *Thought and language*. MIT Press.

Vygotsky, L. S. (1987). Thinking and speech. In *The collected works of L. S. Vygotsky, vol 1.* (pp 37–285) (R. W. Reiber, Ed.; M. J. Hall, Trans.). Plenum.

Vygotsky, L. S. (1998). The crisis at age seven. In *The collected works of L. S. Vygotsky, vol. 5* (pp. 289–296) (R. W. Reiber, Ed.; M. J. Hall, Trans.). Plenum.

Wass, R., & Golding, C. (2014). Sharpening a tool for teaching: The zone of proximal development. *Teaching in Higher Education, 19*(6), 671–684.

Wells, G. (1999). *Dialogic inquiry: Towards a sociocultural practice and theory of education*. Cambridge University Press.

Wertsch, J. V. (1985). *Vygotsky and the social formation of mind*. Harvard University Press.

Zebroski, J. T. (1994). *Thinking through theory: Vygotskian perspectives on the teaching of writing*. Boynton/Cook Publishers, Inc.

# Chapter 4
# Vygotsky and Science Education Research

## 4.1 Introduction

The focus of this chapter is on Vygotsky's concepts and methodologies as they apply to science education research. Science education research has moved historically between three broad theoretical frameworks which governed policy and practice in school science. They are behaviourism, cognitivism, and constructivism (cognitive and social). More contemporary frameworks apply findings from cognitive science and neuroscience research. I start with a very brief overview of these frameworks in the first section.

The second section presents some of Vygotsky's own methodologies, which focus on the development of learning in children and adults. The third section relates to Vygotskian methodology and how it relates to science education research. The fourth and fifth sections consider two specific science education areas, respectively scientific concept formation, and neuroscience, which have become key areas of science education research on formal, informal and science teacher education.

### *4.1.1 Behaviourism*

Behaviourism is based on the principle that scientific learning is a behavioural change which can be induced via appropriate stimuli; it follows the early work of Pavlov (1849–1936), Edward Thorndike (1874–1949), and Burrhus Frederic Skinner (1904–1990), and can be attributed largely to a simple process of stimulus response. The desired response is usually manifest as change in behaviour, because of interaction with the environment via different forms of conditioning. The changes in behaviour are measured as systematic observations. Behavioural learning is the same process in humans and other animals. Much of the conditioning relates to



C. Murphy, *Vygotsky and Science Education*,
https://doi.org/10.1007/978-3-031-05244-6_4

external rewards. In schools, such rewards might include approval from the teacher, a prize of some sort, or good results from tests.

Some strengths of the behavioural approach to science learning and teaching include long-term memory of key facts and information These can be used in both everyday and educational contexts. For example, the 'times tables', often learned at a young age in school, support activities such as shopping, banking, games and quizzes. Science learning is often bolstered using mnemonics, to 'short-cut' retrieval of information, such as colours of the rainbow, the order of geological time periods, and the levels of biological classification.

Limitations of behavioural learning approaches include the lack of learners' characteristics, such as emotion, sociocultural background, and physical characteristics which contribute to their capacity to learn. Extreme behaviourism proposes no need for the 'mind' in learning. In other words, the learner presents as a 'blank slate' and the learning outcome is observed only externally. Behaviourist approaches also seek reductionism and reject complexity in the content of the teaching programmes. Neither is there room for collaboration between learners – behaviourist approaches assume that learning is individual.

### *4.1.2 Cognitivism*

Cognitivism became the dominant paradigm in the 1960s. Cognitivism developed as a theory which criticised behaviourism, in that it did not consider the mind in learning (Clark, 2018). Cognitivism has a focus on the mind as a *black box* which needs to be opened (Ertmer & Newby, 1993). As with behaviourism, cognitivism focuses on individual learning, but highlights the importance of mind. It contends that learning happens in our minds, using processes such as thinking, knowing, memory and problem-solving. It aims at opening the 'black box' of the human mind to find out ways of understanding how people learn. Learned knowledge is considered as schema or symbolic mental constructions, thus developing, and altering such schema defines learning. Cognitivism applies the metaphor of the mind as a computer. Information enters the mind, is processed, and leads to learning outcomes.

### *4.1.3 Constructivism: Cognitive and Social*

Constructivism proposes that learners construct their own understanding and knowledge of the world, through experiencing things and reflecting on those experiences. There are two ways that constructivism is manifest: as cognitive constructivism and social constructivism. In *cognitive constructivism*, it is supposed that children discover scientific concepts because of applying logical thought to results of interaction with objects and phenomena; it is based mostly on the work of Jean Piaget. As with behaviourism and cognitivism, cognitive constructivism assumes that learning is an individual process.

*Social constructivism,* on the other hand, suggests that learning science is bound by, and depends upon, the specific social and cultural context available to the learner. It arises partly from Vygotsky's CHT, based on the premise that there is a complete and constant connection between the individual and society that cannot be deconstructed into analysable elements without losing the characteristics of the whole. It pre-supposes that learning occurs first between people and then internalised within the individual. Detel (2001), in his chapter on social constructivism, suggested that a sociological analysis of science and scientific knowledge is fruitful and reveals the *social nature* of science. Detel (2001) also described three main social constructivist approaches:

- The development of scientific knowledge is seen to be determined by social forces, essentially contingent and independent of rational methods, and analysable in terms of causal processes of belief formation.
- The Edinburgh School of the Sociology of Science maintains that it is not only the development but also the content of scientific theories that is determined by social factors.
- The leading idea of the actor-network theory is that scientific knowledge is an effect of established relations between objects, animals, and humans engaged in scientific practices.

Detel (2001)'s article discusses the historical background of social constructivism and gives a lot more detail on the three approaches mentioned above.

## 4.2 Vygotskian Methodology

Vygotsky was disillusioned with the traditional, objective, biological approach to psychological experimentation on learning, which focused on neural reflexes and impulses, and stimulus-response methods of behaviourism, popular at the time. He objected to the reductive processes of traditional psychology, for instance that the mind could be objectively measured, and that learning was an individual, internal, and innate process, determined in the main by genetics. Instead, Vygotsky's focus was on consciousness, particularly the conscious awareness of learning. Vygotsky conceptualised psychology as a *tool* with which to investigate culture and consciousness, as opposed to a subject of study (Newman & Holzman, 1993). Vygotsky claimed that full consciousness as a human only develops in the presence of other humans, and that learning occurs because of the interrelationship between the learner, other people, and their environment. Thus, the formation of HPFs is a highly complex process in which people interact on the social and environmental planes and internalise these interactions. This transforms their thinking and actions via those interactions which have an emotional connection, with the result of new, more informed, nuanced thinking and learning, which can subsequently lead to logical, agentic, and more productive action.

(i) ***Genetic Experimental Methodology***

Vygotsky rejected the cause-effect stimulus response in favour of a methodology that has emphasis on the *emergent* nature of mind in activity. This methodology is used to study and describe the concept of internalisation in the development of HPFs. He termed this methodology as: *genetic experimental* methodology, in which he and his colleagues studied the *process* of learning, as opposed to the *outcomes*. He examined complex systems in the process of change, using dialectical logic to understand the interrelationships between the components of the system, by using interventions, and by observing the learners' responses, for example:

1. An experimenter observes and reports participants' initial efforts to solve a 'new' problem (intervention) using their existing means.
2. The experimenter then introduces auxiliary means through which the problem can be solved. This mediated assistance is observed and analysed theoretically and methodologically in terms of the learning process.
3. Developmental changes taking place are recorded over several sessions during which learners' appropriate new psychological tools, leading towards the development of HPFs.

A famous example of this methodology is the Vygotsky blocks experiment, described below:

(ii) ***The Vygotsky Blocks Experiment and Concept Formation***

The Vygotsky blocks experiment is one of the most famous experimental approaches in the exploration of concept development using his genetic developmental analysis. Vygotsky and his co-workers explored the process of new concept formation using a series of *double-stimulation* experiments.

The Vygotsky blocks double stimulation experiment was repeated more recently by Towsey (2007, p. 3), who described the blocks as follows:

> *The material comprises 22 wooden blocks of five colors (orange, blue, white, yellow, and green); six geometric shapes (isosceles triangles, squares, circles, hexagons, semi-circles, and trapezoids); two heights; two sizes (diameters); with the labels* ***cev***, ***bik***, ***mur***, *and* ***lag*** *written underneath them (****lag*** *and* ***mur*** *having five blocks each, and* ***cev*** *and* ***bik*** *having six).*

The blocks themselves represent the first stimulus, whilst the abstract labels provide the second, which help to solve the problem. The participants were challenged with sorting the blocks into four coherent groups. Only occasionally was a participant invited to look at a label to see if its clue led towards the solution. The solution is that:

- The *lag* blocks are all tall and big.
- The *mur* are tall and small.
- *Bik* are flat and big.
- *Cev* are small and flat.

Towsey (2007) published a video of the blocks experiment.[1]

The blocks experiment formed part of a series carried out by Vygotsky and co-workers, which led to the proposal of the stages that are passed through in the formation of concepts by children. The stages comprise random grouping, thinking in complexes, pseudoconcepts and true concepts. The youngest children grouped objects 'randomly', according to chance or some other subjective impressions. Older children demonstrated that they were thinking in complexes, in which they began to abstract or isolate different features or attributes. These were related to the child's experience rather than to logical thinking. At this stage the child is showing evidence of organising thoughts, which lays the foundation for more sophisticated generalisations. Such pre-conceptual thinking is deemed necessary for successful mathematics (Berger, 2005) and scientific concept construction. The next stage is the development of pseudoconcepts, which can be confused with true concepts because the learner might be using the right words to describe the concept but lack the logical connections between its parts. The learner can use the pseudoconcept in communication and activities, such as exams, as if it were a true concept. For example, the learner may use the definition of an ionic bond to describe how it differs from a covalent bond without being able to *use* these terms in describing the nature of chemical bonding. The words of the learner and teacher may refer to the same idea, but their meanings may not be the same (Gredler & Claytor Sheilds, 2008). Berger (2005) suggests that true concepts are formed from pseudoconcepts via the *appropriate* use of signs and social (frequently teacher) interventions, thereby forming a bridge between the individual and social meanings. A true concept is bound by *logical* bonds within parts and between different concepts. True concepts are developed with conscious awareness (Gredler & Claytor Sheilds, 2008) and promote the development of everyday concepts into the accepted scientific framework, where they can be used, further developed and critiqued.

(iii) ***Dialectics in Vygotskian Methodology***

Vygotsky was interested in analysing phenomena from different perspectives. Hence, he utilised dialectical analysis which seeks to unify contradictions. He examined mind and matter in their interconnectedness. It is the idea of unifying contradictions which distinguishes it from traditional approaches. For example, in twentieth-century physics, a unified vision of light was as both wave and particle, which led to a broader theoretical understanding (John-Steiner & Mahn, 1996). The same authors describe the unification of the very different elements of hydrogen and oxygen which, when they combine to form water, and which itself goes through transformations from gas to liquid to solid. Vygotsky suggested that dialectical analysis always considers the 'whole' as well as the parts and recognising that the whole is in a state of flux.

[1] A video of this fascinating experiment is available at: https://vimeo.com/groups/chat/videos/10689139

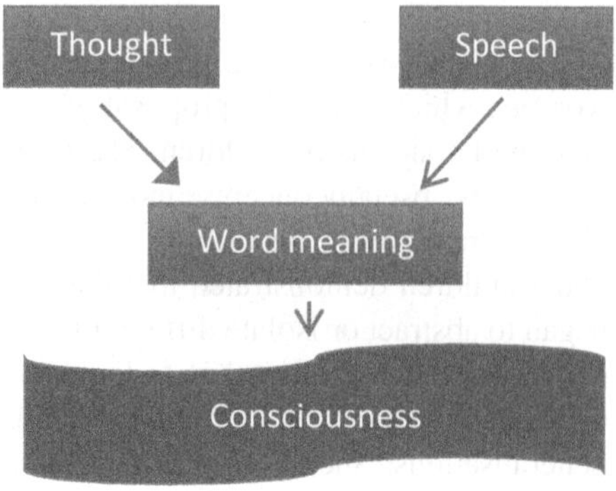


**Fig. 4.1** A schematic representation of a dialectical analysis of consciousness

He investigated the process of how parts relate to a whole, whilst seeing the whole as dynamic, and depending on the parts for its growth/development, for example, thought and speech combine in the process of word meaning, which can be described as the basic 'unit of analysis' (UoA) of consciousness (see Fig. 4.1).

Vygotsky's method includes breaking down a whole into 'units of analysis' which are component parts of the whole, unlike elements which lose the properties of the whole, such as hydrogen and oxygen in the formation of water. Consequently, the UoA of water would be a molecule of water. Tobach (1995) proposed that using a dialectical/cultural-historical approach, theorists can analyse internalisation and individual/social processes as interrelated parts of neurophysiological, psychological, educational, political and cultural systems.

Vygotsky's focus of CHT was the transformative nature of internalisation, which by way of facilitating the development of HPFs in a large mass of people who can think, work and act in ways that can lead towards social transformation. An example, often quoted, is the transformation that occurred in the Soviet Union from the devastation resulting from the First World War, the 1917 Russian Revolution and Stalin's five-year plan, to a nation which put the first artificial Earth satellite into space (Sputnik in 1957), shortly followed by the first man in space (Yuri Gagarin in 1961). Both of these achievements beat the USA, which had none of the devastation on the scale of Russia. Part of this success was attributed to the massive effort in Russia to motivate and educate society to enable such transformation. Vygotsky's work contributed, although some aspects of his research were frowned upon due to his collaboration with European and American colleagues.

Vygotsky's overall aim was to find a theory of human development that would inform pedagogical and rehabilitation practice and overcome developmental defects in 'impaired' and 'abnormal' children, consequently improving human nature (Yasnitsky, 2011). In terms of how to do this, he famously suggested that:

> *The search for method becomes one of the most important problems of the entire enterprise of understanding the uniquely human forms of psychological activity. In this case, the method is simultaneously prerequisite and product, the tool and the result of the study.* (Vygotsky, 1978, p. 65).

Vygotsky maintained that no person is disabled; rather, they are themselves, as they are. It was society that put such people into a single group, called 'disabled'. Vygotsky suggested that each of us depends on specific tools, whether they be crutches, medication, other therapies, vehicles, glasses, wheelchairs, etc., to navigate through our lives.

## 4.3 Vygotsky and Science Education Research Methodologies

In the early part of the twentieth century, learners of science were thought to absorb information derived from lectures, demonstrations and experiments. A study of the history of science education in the USA, carried out by Linn et al. (2016) suggested that behaviourists (e.g., Thorndike, Skinner, Watson, and Pressey) reinforced this transmission view by studying stimulus-response connections and investigating memory and retrieval of information (Thorndike, 1912). Skinner's (1938) work on operant conditioning (building on Watson, 1913) emphasised reward for desired behaviour and inspired programmed learning (Pressey, 1926).

However, John Dewey (1859–1952), as a philosopher, psychologist and educational reformer, distinguished acquiring facts from using the methods of science and called for emphasising scientific reasoning in science instruction (Dewey, 1916). Children's reasoning was a popular area of research at this time; looking back, the two areas of thought that have mostly influenced science education are those from Piaget and Vygotsky.

### *4.3.1 Influence of Piaget and Vygotsky*

Piaget (1930) studied how his own children developed scientific insights and posited a theory featuring developmental constraints. He described four stages of cognitive development, which in summary are:

1. *Sensorimotor stage*: from birth to 2 years, during which infants develop the ability to differentiates self from objects and recognise self as the agent of action, for example, shaking a rattle to make a noise and achieve object permanence.
2. *Preoperational stage*: from 2 to 7 years, during which time children learn to use language and represent objects by images and words, still use egocentric thinking, and can classify objects by a single feature.
3. *Concrete operational stage*: from 7 to 11, when children can think logically about objects and events, achieve conservation of number and volume, classify objects according to several features, and order them in series.

4. *Formal operational stage*: from 11 years onwards, when they can think logically about abstract propositions and test hypotheses systematically, and become concerned with the hypothetical, the future and ideological problems.

Piaget described the process of cognitive development as comprising the mechanisms of assimilation, accommodation and equilibration. Assimilation is the process in which a learner encounters a new idea, and "fits" that idea into what they already know. Accommodation of knowledge is more substantial, such that they alter their existing ideas, or schema, to accommodate this new information in the production of new, more informed schema. The driver of this process of assimilation and accommodation is equilibration. Equilibration involves learners striking a balance between themselves and the environment, between assimilation and accommodation. When they experience a new event, disequilibrium sets in until they can assimilate and accommodate the new information and thus attain equilibrium. There are many types of equilibrium between assimilation and accommodation that vary with the levels of development and the problems to be solved.

Piaget's research methods relied predominantly upon clinical methods, using probing questions. He was interested in the *errors* children made, and he searched for a systematic pattern in the production of children's errors. He developed his theory of genetic epistemology from his studies on how knowledge is acquired and developed. Piaget believed that intelligence arises progressively in the baby's repetitive activities. His theory states that the precursors of thinking and language lie in the elementary actions, perceptions and imitations of babies. His work was influenced strongly by evolutionary theory: a child has to 'adapt' to an environment by altering cognitive structures. He emphasised internal, self-directed, individualist development, and was famous for his promotion of 'discovery learning'. In his own words:

> *Children should be able to do their own experimenting… In order for a child to understand something, he must construct it himself, he must re-invent it.* ***Every time we teach a child something, we keep him from inventing it himself.*** (Piaget et al. (1972), p. 27, emphasis added)

Another frequently quoted citation from Piaget's work emphasises the discovery learning approach, particularly in science, which formed the basis of UK education policy in the 1960s, via the advice given in the *Plowden Report* (1967) drawn from Piaget's *oeuvre*, such as:

> *Spontaneous development is what the child learns by himself, what none can teach him, and he must discover alone…* (Piaget, 1973, p. 2).

And the advice for science learning in primary schools suggested:

> *… treatment of the subject matter may be summarised in the phrase 'learning by discovery'. In a number of ways, it resembles the best modern university practice. Initial curiosity, often stimulated by the environment the teacher provides, leads to questions and to a consideration of what questions it is sensible to ask and how to find the answers.* (Plowden Report, 1967, p. 242)

However, the *Plowden Report*'s focus on discovery learning in science, had some problematic consequences. In a *Times Education Supplement* editorial, Colin Richardson (1997) looked back on some of its impact. Whilst he welcomed the positive tone of the report, he commented on the difficulty in implementing its key focus on individual learning, particularly in large classes, which denied too many children sustained interaction with the teacher and other students. Another effect highlighted by Richardson was of an over-emphasis on classroom layout, displays, etc., and insufficient attention to curriculum content. Another positive factor was that for a few years following the report, primary education was considered important and primary teachers were more valued, but that context was ephemeral and there was little evidence of it returning. Science learning in primary schools today in the UK and Ireland, and probably most other countries, follows a content-driven curriculum.

Piaget's work, according to many researchers, particularly in relation to his stages of cognitive operations, underestimates what children can do. For example, Brown (1994, p. 10) suggested that following these stages:

> *encourages sensitivity to what children of a certain age* ***cannot*** *do because they have not yet reached a certain stage of cognitive operations.* (Emphasis in original)

Essentially, Piaget's thesis centres on the idea that development precedes learning. Vygotsky was born in the same year as Piaget, in 1896, but his life was much shorter. Piaget lived until he was 84 years, whilst Vygotsky died when he was only 37 years old. The research of Vygotsky and Piaget shows similarities and differences. A key difference is the focus on the learner as individualist constructivist (Piaget) or social constructivist (Vygotsky). Piaget's cognitive developmental stages imply that development leads learning, as children are not capable of carrying out certain tasks, for example, conservation of number and volume, until they have reached the concrete operational stage. On the other hand, Vygotsky's ZDP suggests that learning can be enhanced by providing conditions which promote learning beyond their actual level of development, towards their potential level of development, by way of functions which are already developing but not quite mature. In his Vygotsky's words (1978, p 86). the ZPD comprises: "functions which have not yet matured but are in the process of maturing... 'buds' or 'flowers' of development rather than 'fruits' of development", and he proposed that it represents "the domain of transitions that are accessible by the child" (Vygotsky, 1987, p 211).

Vygotsky first described the creation of ZPDs in terms of levels of assistance given to learners of the same 'actual' cognitive level (as can be measured, for example, from same IQ scores) in solving more difficult tasks. Despite being measured at the same level, one child might solve the task with very little help (the support is thus within their ZPD for the task), whilst another may not solve it even after several different interventions designed to support the learning (the support is outside the learner's ZPD for this task). Such interventions may include demonstrating the problem solution to see if the child can begin to solve it; starting to solve it and asking the child to complete it; asking the child to solve the problem with the help of a child who is deemed more 'able'; explaining the principle of the needed solution;

asking leading questions; analysing the problem with the child, etc. (Gredler & Claytor Sheilds, 2008). Vygotsky considered performance on summative tests as an indication of the child's past knowledge and argued that "instruction must be orientated towards the future, not the past" (Vygotsky, 1962, p. 104).

Vygotsky's usage of the ZPD can be applied to support and optimise students' learning and achievement. By using the ZPD concept, teachers, family, and/or fellow learners can help to create conditions which make the learner want to learn for the sake of learning, as opposed to external rewards. Elena Kravtsova was Vygotsky's granddaughter, who sadly died in 2020. She carried out research and development using Vygotsky's CHT. She gave an example of the ZPD within CHT concerning Russian immigrants who had lived in Germany for 20 years without learning German. A teenager amongst them watched German cartoons, played clips of interest repeatedly, and hence learned the German required to enjoy the cartoons. This level of German 'incidental' learning enabled him to use the language much more extensively (Kravtsova, 2017). In this context, the main goal of learning German was to create conditions for developing a person's ability to be the subject of their own behaviour, activity and cognition. The key idea for teachers is to develop and provide ZPDs for the learners, such that they can learn science to solve relevant and interesting problems, enhance their experience of science lessons, and support their further learning in the area.

## 4.4 Research on Scientific Concept Development

Scientific concept development is considered in the context of secondary-level science education in this Chap. 7 of this book. It focuses on concept development research and practice. The process for developing scientific concepts has been debated for many years. One theoretical construct, conceptual change, became very popular in explaining how scientific concept development took place, although it is now considered less important. It was believed that students suffered 'misconceptions' about scientific phenomena. These 'misconceptions' were seen to arise from 'everyday' thinking. The notion of conceptual change suggested that when challenged with the 'correct' scientific concept, students went through a process of cognitive conflict, in which they battled with the 'everyday' and scientific explanations, and, once they accepted the scientific explanation, they were said to have experienced a process of conceptual change, drawing on their growing science knowledge and that of teachers and peers (Hewson et al., 1998). Unfortunately, it was frequently observed that students reverted to the everyday concept of scientific phenomena.

The reversion to everyday thinking was evidenced in a video produced by the Harvard-Smithsonian Center for Astrophysics in 1987. In its famous opening scene, graduating Harvard seniors (one of whom had taken several physics classes) were asked to explain what causes the seasons and the phases of the moon. Some students suggested that eccentricity in Earth's orbit made Earth warmer when it was closest to the sun. Some thought that the phases of the moon were caused by Earth's shadow. Once seeing their ideas and the scientific explanations tested, students tended to accept the new ideas they were being taught but would occasionally try to blend the old and new ideas or revert to the old ideas entirely. The film-makers suggested that even the brightest students in the class had false ideas, which were based on enduring misconceptions that traditional instructional methods cannot overcome. Conceptual change, therefore, has not delivered the learning gains needed. Many university physics students still misunderstand very basic concepts, as evidenced by consistent poor performance on the Force Concept Inventory (FCI) test (Miller et al., 2013). Miller et al. looked to different explanations as to why some scientifically wrong ideas persist at university. They suggest that unless learners are *using* scientific concepts in, for example, problem-solving, it is unlikely that they will retain the scientific explanations they are presented with after they have learned them for a test or examination. It could be that such scientific concepts are not good for learning out of context and that different pedagogical approaches are needed to improve students' understanding of specific concepts.

Another factor that is becoming increasingly important in science conceptual learning is the role emotion plays. Vygotsky proposed the importance of the unity of affect and intellect in the ZPD – that emotion and learning are interdependent. It is impossible to learn without emotive engagement (Reid, 1788/1969) and other higher mental functions (Mahn & John-Steiner, 2002). Matthews (2015) provides more discussion of the crucial role of emotion for science learning.

Research into scientific concept development in children and students over the past century has identified the prevalence of pseudoconcepts in science learning: students can repeat the scientifically correct concept to gain examination credit without understanding its nature or being able to apply it.

### 4.4.1 Concept Development Is Not a Linear Process

Vygotsky (1986) proposed a dialectical, as opposed to a linear, model for the development of scientific concepts: "the child's scientific and [her or] his spontaneous [everyday] concepts… *develop in reverse directions* (original italics) …they move to meet each other" (p. 192). For example, the students' everyday concept of a pool of water develops more scientifically when they learn about evaporation; at the same time, their concept of *evaporation* will become more every day to them when applied to familiar contexts, such as pools of water and perspiration.

Vygotsky proposed that teachers can create a ZPD between the scientific and everyday concepts by illustrating and emphasising the relationships between them and showing how the scientific concept can be utilised to explain the everyday concept, whilst simultaneously raising the everyday concept towards its scientific conceptualisation (see Fig. 3.1, Chap. 3). For instance, a child may have a rich understanding of the everyday concept brother but not be able to define it in the more logical, conceptual way as *male sibling* (Panofsky et al., 1990). The task of the teacher, for Vygotsky, is not to evaluate individual conceptions as correct or as 'misconceptions', but rather to help the child, through instruction with respect to the relationship between concepts within a system of concepts, and to develop conscious awareness and voluntary control of their own thinking (Wells, 2008).

### *4.4.2 The Process of Forming Scientific Concepts*

Most of the science education research literature regarding scientific concept formation relies on the work of Piaget (1955, 2001) and Vygotsky (1986) and their followers. The general argument is whether (a) it occurs via the replacement of a children's (egocentric) concepts by adult ones as children get older (Piaget) or whether (b) scientific concepts are formed from learned experiences in which children first exhibit 'pre-conceptual thinking', which shows evidence of organising thoughts and some abstraction, but not of systematic thinking or sophisticated abstraction, and later develop scientific concepts via the interplay between the 'everyday' and 'scientific concepts', using these concepts to help explain scientific phenomena (Vygotsky).

Murphy (1987) considered that both explanations underestimated very young children's thought in terms of coherence and systematic thinking. Her study involved 280 children (5–7-year-olds) who were recorded during a game in which a child described the meaning of a scientific word (concept) without using it for the rest of the class to guess. Despite a large proportion of the responses being context-bound, children's descriptions evidenced definite, coherent ideas about most of the concepts. Many 7-year-olds demonstrated a level of abstraction beyond that predicted by Piaget or Vygotsky. For example, the description of amount as degrees; weather as a sort of condition; transport as types of vehicles, and idea as a plan. She argued that the limiting factors in the development of scientific concepts could be largely related to lack of vocabulary, experience and specific conceptual frameworks, as opposed to a lack of systematic thinking.

### *4.4.3 Vygotsky and Classroom Research in Scientific Concept Development*

In teaching scientific concepts, the aim is to develop student understanding of the concept(s), make the learning meaningful in relation to school and out-of-school experiences, and to enable students to appreciate and *interact* with the world of science, which uses concepts in codified and regulated contexts. Another aim, not as prevalent in the literature, is to engage students to think and learn about the world as it is and its future in science lessons. Below are some classroom research examples of how scientific concepts can be used in more interesting, dynamic, and active ways in learning and teaching science.

### 4.4.4 Creating ZPDs to Bridge Inquiry-Based Science Education (IBSE) with Scientific Convention

IBSE needs to connect the task explicitly with the scientific context to ensure that the learning is meaningful. Rubtsov (2007, p. 10) described such a setting:

> *Two children must work together to balance a set of weights on a calibrated arm by moving, adding, or removing weights. To solve this problem, they must consider the relationship between each weight and its distance from the arm's centre of gravity. One participant is allowed to move the weights along the arm but not to add or remove weights; the other may increase or reduce the number of weights, but not move them. This division of activities, therefore, requires the two participants to work together, coordinating their activities to solve the task successfully. As the children move to the next problem, they switch roles.*

Rubtsov (2007) cautioned that such activities, whilst promoting reflective thinking, do *not* guarantee that each pupil will be able to identify the essential elements of the task. He suggested that to increase the effectiveness of the activity, pupils should be provided with pictorial and symbolic models to represent the problems they are solving and the steps they use to solve them. Hence, they will be applying a conceptual framework into which their activity can be made scientifically meaningful. The pictorial and symbolic models, together with the discussion generated between learners as they complete the task, will become more meaningful to the children.

This type of work helps to promote thinking and stimulate children to reflect and explain to understand how their experiences, and their context-bound knowledge fit into a larger scientific system (Howe, 1996). The teacher is essential here to guide the work and provide the conceptual framework. Howe argues that a contrasting, Piagetian approach would prefer that the children worked on their activity without teacher intervention. Anne Howe (1996, p. 46) maintained that:

> *decontextualized tasks, chosen to represent a process but unrelated to children's everyday knowledge or interests, would not have a place in a science curriculum informed by a Vygotskian perspective.*

Bereiter (1994) argued that school science exists predominantly in a context of mere learning, and that teachers should aim to move it towards one which constitutes knowledge building. He suggested that this situation could be achieved by activities aimed at collaborative creation in the classroom. Knowledge-building also requires that students use the scientific 'tools', including signs and symbols, so that their ideas can translate easily into scientific contexts.

## 4.5 Vygotsky and Neuroscience

Neuroscience comprises the study of the human nervous system, the brain, consciousness, perception, memory, and learning. Research on human reasoning informed by the emergence of the first transistorised computers in the 1950s provided cognitivists with a helpful analogy for the human mind as an information processor. This view of thinking and human behaviour remains popular today but is being challenged by many scientists in different disciplines, for example, the Nobel Prize-winning neurobiologist, Edelman (2017), computer scientist Rapaport (2018), sports scientists Raiola and Di Tore (2017) and philosopher Erden (2017).

When educators take neuroscience into account, they organise a curriculum around real experiences and integrated, 'whole' ideas. They focus on instruction that promotes complex thinking and the 'growth' of the brain. Neuroscience proponents advocate continued learning and intellectual development throughout adulthood.

The nervous system and the brain are the physical foundation of the human learning process. Neuroscience links observations about cognitive behaviour with the actual physical processes that support such behaviour. The science is still 'young' and is undergoing rapid, controversial development. Some of the key findings from neuroscience and learning indicate that the brain is not a computer that requires linear and/or parallel-type connections. It is more akin to a self-organising system of billions of neurons, which are not connected in a linear fashion. Instead, the connections are loose, overlapping, webbed and flexible, and some are redundant. The brain also changes over time.

In essence, for learning to occur, sensory cells (e.g., in the eyes and ears) receive external stimuli (e.g., light, or cold), which is converted to electrical impulses which travel to the brain as 'messages'. The brain cells (neurons) 'interpret' these messages and send responses to parts of the body (e.g., the eyes) which then respond (e.g., they see). Specific external stimuli will determine which connections are made and therefore which messages are transmitted. Transmission is strongly affected by chemicals (neurotransmitters) which transmit nerve impulses: tranquilisers modify chemicals in the synapse (gap) between neurons and affect the speed of transmission, and other such as lysergic acid (LSD) alter the balance between different neurotransmitters. Each neuron has tens of thousands of connections to others, forming very complex 'neural pathways', which link together to form neural networks. *Nobody yet knows for sure how individual firings of neurons can lead to memory, pattern recognition, logical reasoning, emotion, and consciousness***.** The brain also changes with use, *throughout the lifetime*.

Many neuroscientists define learning as changing the structure and actions of neurons so that they hold information in the long-term memory, situated in the temporal and parietal lobes of the cortex. Learning occurs via changes in the amounts of different neurotransmitters that neurons produce and changing the number and the nature of the connections between neurons. Mental concentration and effort alter the physical structure of the brain. There are tens of billions of neurons and about 1000 trillion connections. As the brain is used, certain patterns of connection are developed, making each connection easier to create next time. This is how memory develops.

Neuroscience research has found five promoter mechanisms whereby short-term learning changes into long term learning[2] which change in number of neurotransmitters and neural connections.[2] Such promoters are:

- Innate learning programmes.
- Repetition of information.
- Excitement at the time of learning.
- Aating carbohydrates at time of learning.
- Taking 8–9 h of sleep after learning.

Educators who take neuroscience research findings into account tend to organise a curriculum around real experiences and integrate 'whole' ideas. They focus on instruction that promotes higher level complex thinking and the 'growth' of the brain. Neuroscience proponents advocate continued learning and intellectual development throughout adulthood. Many researchers in neuroscience have discovered the close links between Vygotsky's ideas about learning and development with their findings. For example, (Tuomi, 2012) acknowledged that Vygotskian theory can be taken into consideration when dealing with the thorny question of how the passages of impulses between neurons lead to learning. Vygotsky claimed that HPFs in humans are products of cultural and historical development. Child and adult learners respond to signs, signals and historically accumulated systems, of which language is the most important, in developing cognition (Luria & Vygotsky, 1992). Thus, perception is mediated by cultural tools and artefacts, which determine human responses.

One current example of culturally mediated learning is the difficulty many young children now face when introduced to the use of pens and pencils in school. It used to be the case that parents, carers, older siblings and other children were in contact with pens, pencils, etc., from birth, as they would be constantly modelling writing, such as grocery lists, birthdays cards, etc., and writing letters. Many children now grow up in an environment in which there are very few, if any, pens, or pencils lying around. Almost no adult is modelling their use; instead, they are using laptops and iPads, other tablets, and smartphones. The children play with these devices, which leads to a lack of development of the muscles required for holding, writing, and drawing with pens and pencils. Another issue for young and older children in science learning arises when schooling is separated from the cultural context in which they have been living, commonly experienced when children immigrate to a new country.

Cognitive development relies on cultural mediation, modelled by other humans. Thus, people from different cultural environments cannot be expected to access the same tools and signals. In westernised countries, beginning in preschool, children

[2] Online slides, available at: https://slideplayer.com/slide/9155994/

are often segregated from adults and receive culturally important information and instruction outside the context of skilled activities. On the other hand, in many tribal cultures, children are in close contact with adults for most of the day and observe and interact with adults while they perform culturally important activities.

The experiments of Vygotsky and Luria in Uzbekistan in the early 1930s, where they compared different forms of cognition between farmers who had never attended formal schooling with people who had experienced formal schooling, demonstrated that the groups used different cultural tools and exhibited different forms of cognition. Those who retained a traditional, non-literate culture and way of life tended to solve problems by using functional reasoning, reflecting their everyday life practical experience. They rejected more abstract ways of problem-solving, such as the use of classification and generalisation. However, people who had received some form of formal education demonstrated a clear preference for the verbal-logical form of problem-solving. These results were confirmed several times in other studies. As recently as 2018, Glozman reproduced Luria and Vygotsky's research, using the same tests on classification and generalisation, and the same neuropsychological battery and projective drawings on life attitudes (Glozman, 2018). This study, conducted in the north of the Kamchatka Peninsula, in Russia, compared the cultural tools and cognitive functions used by settled groups of people with nomadic herdsmen and women in the tundra. Their results showed that nomadic and settled groups with the same level of schooling (most attended primary school) differed in neuropsychological tests, which revealed the influence of their respective social life conditions.

## 4.6 Summary and Conclusion

This chapter has focused on aspects of both Vygotsky's own research and the use of his work in science education contexts. It has provided ideas and experimentation from Vygotsky's work that led him to discover ways to advance development via learning in the ZPD. His work showed how the dialectical conception of scientific concept development facilitates teaching concepts towards 'true' understanding, as opposed to the too-often situation that students are only learning to the pseudoconcept level (in which they can define a concept and use the right words in the right places, but they cannot use that concept or link it to other, related ones). His dialectical analysis of consciousness led to his identification of the unit of the UoA, [the smallest part which represents the whole, such as a molecule of water as the UoA of water] of consciousness, the UoA is word meaning, which is developed from thought and speech. Vygotsky's work on social interaction as the basis of learning is key to collaboration studies.

Chapter 4 has completed Part II of this book, which introduced Vygotsky's influence on science learning, teaching and research. Part III applies this work to science learning and teaching in formal school and other learning institutions, including teacher learning in science pedagogy.

## References


Bereiter, C. (1994). Constructivism, socioculturalism and Popper's world 3. *Educational Researcher, 23*, 21–23.

Berger, M. (2005). Vygotsky's theory of concept formation and mathematics education. In H. L. Chick & J. L. Vincent (Eds.), *Proceedings of the 29th conference of the International Group for the Psychology of Mathematics Education* (pp. 153–160). University of Melbourne.

Brown, A. L. (1994). The advancement of learning. *Educational Researcher, 23*(8), 4–12.

Clark, K. R. (2018). Learning theories: Cognitivism. *Radiologic Technology, 90*(2), 176–179. PMID: 30420575.

Detel, W. (2001). Social constructivism. In *International encyclopedia of the social & behavioral sciences* (pp. 14264–14267). Elsevier.

Dewey, J. (1916). Method in science teaching. *General Science Quarterly, 1*(1), 3–9.

Edelman, G. M. (2017). Building a picture of the brain. *Annals of the New York Academy of Sciences, 882*(1), 68–89.

Erden, Y. J. (2017). Being "simple-minded": Models, maps and metaphors and why the brain is not a computer. In *Representation and reality: Humans, animals and machines*. Springer.

Ertmer, P. A., & Newby, T. J. (1993). Behaviourism, cognitivism, constructivism: Comparing critical features from an instructional design perspective. *Performance Improvement Quarterly, 6*(4), 50–72.

Glozman, J. M. (2018). A reproduction of Luria's expedition to Central Asia. *Psychology in Russia: State of the Art, 11*(2), 4–13.

Gredler, M., & Claytor Sheilds, C. (2008). *Vygotsky's legacy: A foundation for research and practice*. Guildford.

Hewson, P. W., Beeth, M. E., & Thorley, N. R. (1998). Teaching for conceptual change. In K. G. Tobin & B. J. Fraser (Eds.), *International handbook of science education* (pp. 199–218). Kluwer.

Howe, A. (1996). Development on science concepts with a Vygotskian framework. *Science Education, 80*(1), 35–51.

John-Steiner, V., & Mahn, H. (1996). Sociocultural approaches to learning and development: A Vygotskian framework. *Educational Psychologist, 31*(3–4), 191–206.

Kravtsova, E. E. (2017). The sense and the meaning of cultural-historical theory of L. S. Vygotsky. *Revue Internationale du CRIRES: Innover dans la Tradition de Vygotsky, 4*(1), 35–47.

Linn, M. C., Gerard, L., Matuk, C., & McElhaney, K. W. (2016). Science education: From separation to integration. *Review of Research in Education, 40*(1), 529–587.

Luria, A. R., & Vygotsky, L. S. (1992). *Ape, primitive man, and child essays in the history of behaviour*. CRC Press.

Mahn, H., & John-Steiner, V. (2002). The gift of confidence: A Vygotskian view of emotions. In G. Wells & G. Claxton (Eds.), *Learning for life in the 21st century* (pp. 46–58). Basil Blackwell Publishers Ltd..

Matthews, B. (2015). The elephant in the room: Emotional literacy/intelligence, science education, and gender. In D. Corrigan, C. Buntting, J. Dillon, A. Jones, & R. Gunstone (Eds.), *The future in learning science: What's in it for the learner?* (pp. 193–212). Springer.

Miller, K. A., Lasry, N., Chu, K., & Mazur, E. (2013). Role of physics lecture demonstrations in conceptual learning. *Physical Review Special Topics – Physics Education Research, 9*(2), 1–5.

Murphy, C. (1987). Primary science: An analysis of concept formation in 5-to 7-year-old children. Unpublished M.Ed. thesis, Queen's University Belfast.

Newman, F., & Holzman, L. (1993). *Lev Vygotsky: Revolutionary scientist*. Routledge.

Panofsky, C., John-Steiner, V., & Blackwell, P. (1990). The development of scientific concepts and discourse. In L. C. Moll (Ed.), *Vygotsky and education: Instructional applications of sociohistorical psychology*. Cambridge University Press.

Piaget, J. (1930). *The child's conception of physical causality* [La causalite physique chez l'enfant]. Kegan Paul.

Piaget, J. (1955). *The language and thought of the child*. New American Library.

Piaget, J. (1973). *To understand is to invent: The future of education*. Grossman Publishers.

Piaget, J. (2001). *Studies in reflecting abstraction*. Psychology Press.

Piaget, J., Wolff, P. H., Spitz, R. A., Lorenz, K., Murphy, L. B., & Erickson E. H. (1972). *Play and development: A symposium with contributions* (M. W. Piers, Ed.). W. W. Norton & Company.

Pressey, S. L. (1926). A simple apparatus which gives tests and scores—And teaches. *School and Society, 23*, 373–376.

Raiola, G., & Di Tore, P. A. (2017). Motor learning in sports science: Different theoretical frameworks for different teaching methods. *Sport Science, 10*(Suppl. 1), 50–56.

Rapaport, W. J. (2018). What is a computer? A survey. *Minds and Machines, 28*(3), 385–426.

Reid, T. (1788/1969). *Essays on the active powers of the human mind*. MIT.

Richardson, C. (1997). Plowden report revisited: Individuality, equality and discovery. *Times Education Supplement Editorial*. Retrieved from http://www.tes.com/news/polwden-report-revisited-individuality-equality-discovery

Rubtsov, V. (2007). Making shared learning work. In *Children in Europe: Vygotsky edition* (pp. 12–13). Children in Scotland.

Skinner, B. F. (1938). *The behaviour of organisms: An experimental analysis*. Appleton-Century.

The Plowden Report. (1967). *Children and their Primary Schools*. A Report of the Central Advisory Council for Education (England). Retrieved from: http://www.educationengland.org.uk/documents/plowden/plowden1967-1.html

Thorndike, E. L. (1912). *Education: A first book*. Macmillan.

Tobach, E. (1995). Comments on the present status of comparative psychology. *Polish Psychological Bulletin, 26*, 203–229.

Towsey, P. M. (2007). *The magic lantern presentation of the illustrated companion to chapter 5 of thought and language* (1986). University of the Witwatersrand. Retrieved from: https://link.springer.com/chapter/10.1007/978-3-319-16543-1_10

Tuomi, I. (2012). Educational neurosciences–More problems than promise? Retrieved from: http://citeseerx.ist.psu.edu/viewdoc/download?doi=10.1.1.694.1490&rep=rep1&type=pdf

Vygotsky, L. S. (1987). Thinking and speech. In *The collected works of L. S. Vygotsky, vol. 1* (pp. 37–285) (R. W. Reiber, Ed.; M. J. Hall, Trans.). Plenum.

Vygotsky, L. S. (1962). *Thought and language*. MIT.

Vygotsky, L. S. (1978). *Mind in society: The development of higher psychological processes* (M. Cole, V. John-Steiner, S. Scribner, & E. Souberman, Eds.). Harvard University Press.

Vygotsky, L. S. (1986). *Thought and language*. MIT Press.

Watson, J. B. (1913). Psychology as the behaviourist views it. *Psychological Review, 20*, 158–177.

Wells, G. (2008). Learning to use science concepts. *Cultural Studies in Science Education, 3*(2), 329–350.

Yasnitsky, A. (2011). Lev Vygotsky: Philologist and defectologist. A socio-intellectual biography. *Portraits of pioneers in developmental psychology, 7*, 105–129.

# Part III
# Application of Vygotskian Theory in Science Education Contexts

# Chapter 5
# Vygotsky and Science Learning in the Early Years

## 5.1 Introduction

This chapter focuses on children's science learning in the early years from a cultural-historical perspective. It draws on the work of key researchers in Vygotskian early childhood science, for example, Fleer (2012), Edwards and Bird (2017), Hao and Fleer (2016) and Kravtsov (2010).

The Vygotskian focus is on play and imagination in science learning. The chapter reflects on the interrelationship between development and learning, linking home and school experiences, playful science learning, digital learning in the early years, and the orientation of young children in terms of time, place and substance (materials). I have included my observations and experiences of playful learning in Vygotsky-inspired schools in Russia, and research on adopting Vygotskian principles into early years learning in an Irish school, which will be described (see Doherty, 2013).

There are six sections to the chapter. After the introductory Sects. 5.1 and 5.2 gives an overview of the role of play and imagination in early years science learning. This is followed by Sect. 5.3, which describes how play develops during the early childhood years. Section 5.4 describes how children's playful experiences set the foundations of their ideas relating to scientific method. Section 5.5 is the role of play in early scientific concept development, and Sect. 5.6 discusses play and digital learning. The chapter concludes with implications and ideas for effective science learning and teaching in early years education contexts.


C. Murphy, *Vygotsky and Science Education*,
https://doi.org/10.1007/978-3-031-05244-6_5

## 5.2 Play and Imagination in Early Years Science Learning

Play and imagination are key elements of children's learning in all areas. Historically, until the eighteenth century, children were seen as miniature adults, and there was no concept of childhood. During the Industrial Revolution, children worked as labourers in factories, and it was only in 1833 that children under the age of 9 were no longer allowed to work in factories in Britain. Too many children were dying from the work or from fatal accidents in the factories. Child labour was finally abolished in most countries by the early twentieth century and replaced by mass schooling to prepare children to work in factories when they were older. At this time, childhood became considered as a 'phenomenon', and research into children's development and learning was established. The studies of classical play theorists, such as Jean-Jacques Rousseau (1712–1778), Frederick Froebel (1782–1852) and Dewey (1859–1952), were key to the development of a dramatic change in societal views and attitudes towards children. A strongly held belief that play was critical to children's learning and development emerged.

These early theorists were strong advocates for children's learning in and from nature as active learners, suggesting that children learned best when they were given opportunities to observe and interact with nature and life. The work of Froebel, known as the creator of the first kindergarten, was highly significant at the time, and endures today. Froebel proposed that children learned through their play, leading towards a life in harmony with others and with nature. His approach emphasised the importance of self-chosen activities in their play. He encouraged teachers to build on and guide children's play within their own worlds. Froebel's ideas fit well with a Vygotskian cultural-historical view of children's learning, in which a child creates an imaginary situation.

Very young children play with neutral objects; for example, a stick becomes a 'horse'. The child will use the stick to trot, gallop and jump like a horse, until it is no longer good enough to satisfy them – they will look around for a better one. Or they might play with an empty box, which becomes a house, garage, shop, etc. All this time they are abstracting from the neutral object to an imagined situation, thus developing symbolic thought. Such behaviour and abstraction are observable in children as young as 1 year of age. They are also constructing their theories about the world, raising, and testing their hypotheses, and revising their views based on evidence.

As they get a little older, children engage more in group play, during which they invent rules of behaviour and learn to self-regulate. Young children usually play 'above' themselves, in a way which creates a zone for development (ZPD), as they use their imagination to act out behaviours which bridge the imagined and real situations. If teachers encourage children and adults to role play scientific contexts, they are encouraging them to use their imagination in ways which can lead to both true concept development and novel ideas for research. A quote often attributed to Einstein is that "Play is the highest form of research." The actual quote is from Scarfe (1962):

*The desire to arrive finally at logically connected concepts is the emotional basis of a vague play with basic ideas. This combinatory or associative play seems to be the essential feature in productive thought.*

Role-playing scientific concepts is also fruitful, such as students acting out mitosis, or muscle contraction. Such embodiment of complex processes facilitates understanding as the ideas are acted out in collaboration, via discussion between students to perform an accurate representation of the concept. Other ideas for role paly include students 'playing' at being molecules to a metronome beat, used to represent heat – students discover that molecules need space when heated, and report improved understanding, explanation, and interpretation of the concept of heat.

Imagination and play comprise a vehicle for the development of important insights, perhaps leading to the construction of complete concept models. It could be argued that *for young children, play is imagination in action*, whereas in *adults and older children, imagination is play without action*. Scientists 'playing' with materials and/or data need to draw on an imaginary situation and rules together to infer creatively how what they imagine can be related to real life.

The psychiatrist and psychoanalyst Carl Jung (1875–1961) suggested that:

*The creation of something new is not accomplished by the intellect, but by the play instinct acting from inner necessity. The creative mind plays with the objects it loves.* (Carl Jung, as cited in Ackerman, 1999)

## *5.2.1 Cultural-Historical Conceptualisation of Play*

A cultural-historical conceptualisation of play draws upon the importance of cultural experiences and social interactions in play activities. This view contrasts with a more Piagetian view which considers an 'average child' and that play is based on the maturation of internal abilities. In contrast, a cultural-historical perspective suggests that in play, children build diversity and complexity from historical and/or social experiences (Hao & Fleer, 2016). Hedegaard (2009, p. 65) suggests that there are three perspectives within a cultural-historical theoretical framework, all three of which impact on children's play:

(a) society's perspective with traditions that implies values, norms, and discourses about child development.
(b) different institutions' perspectives that include different practices; and
(c) children's perspectives that include their experiences, engagements, and motivations.

Hao and Fleer (2016) add the cultural-historical importance of *sign* as embedded in collective imaginary play when a child's scientific learning is considered. Fleer (2012) proposed that science itself is a "collectively imagined social construction that is shared through others" (p. 34) such as scientists, teachers, and students. Thus, early learning, play, imagination and creativity are all fostered by the richness and variety of previous experiences, from which children can construct products of fantasy. The connectedness across contexts and time emphasises the importance of recognising and leveraging the broader ecology of learning in early childhood (Hayes et al., 2017; Bronfenbrenner & Morris, 2006) and attending to the multiple, localised funds of knowledge available to children from their day-to-day life experiences.

The legacy of the early theorists and a cultural-historical approach to play has influenced the now common perception of play as holistic, and a key part of the children's worlds. Play should be as well-resourced as possible, with rich social experiences for children, as well as the environment in which they play at home and school. Vygotsky advocated enhancing strong links between home and school (e.g., two-way communication, as opposed to the more common one way, from school to home). In Vygotsky's words:

> *Ultimately only life educates, and the deeper that life, the real world, burrows into the school, the more dynamic and the more robust will be the educational process.* ***That the school has been locked away and walled in as if by a tall fence from life itself has been its greatest failing.*** *Education is just as meaningless outside the real world as is a fire without oxygen, or as is breathing in a vacuum. The teacher's educational work, therefore, must inevitably relate to his creative, social and life work. Only he who exerts a creative role in real life can aspire to a creative role in pedagogics. It is just for this reason, that in the future the educator will also be an active participant in society.* (Vygotsky, 1997, pp. 345–346, emphasis is mine)

## 5.3 Development of Play During the Early Years

Research and theory development on play in the Vygotsky Institute in Moscow took Vygotsky's ideas about the importance of play and constructed a sophisticated understanding of its development and how it supports a child's learning and development in the early years (Kravtsova & Maximov, 2014).

The earliest stage, according to Kravtsova and Maximov (see Fig. 5.1), is Object Manipulation, during which a child plays with objects, such as sticks, which might represent different imaginary characters, for example, a long stick could represent a horse in such play. The second stage, known as *Director's Play*, is characterised by a child using several different objects to create an imaginary situation or story. During the third stage, *Image Play*, a child mimics experiences that they have observed in real life, or those on television, or from stories that they have heard. The child will 'become' the person, animal, or other character by creating the image using dressing-up, mimic the language and tone of the voice(s) heard, and copy the behaviour. The fourth stage, *Subject-Plot Role Play*, is more sophisticated, in which the child is the subject of the play, not simply creating a perfect image, but acting out the character as themselves in the role of doctor, scientist, frog, patient, playing their own way of being in the role. The child (or children) become characters and

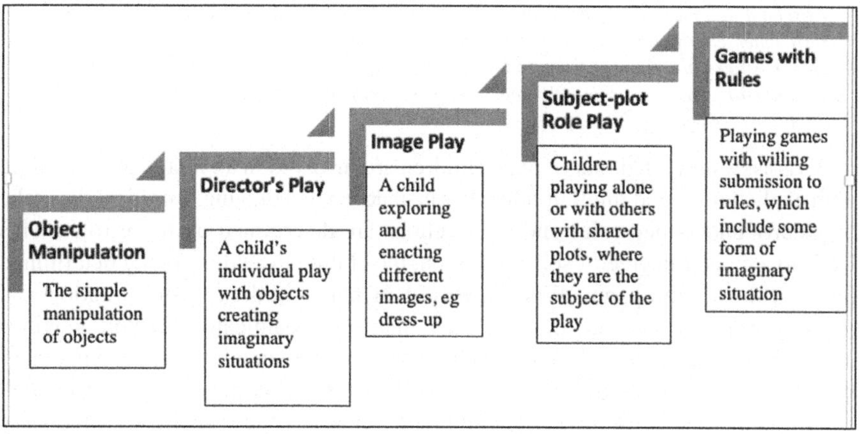


**Fig. 5.1** Summary of the development of play during the early years

develop the role(s) by incorporating aspects of their own personality into it and reacting in their own way to certain events. The fifth stage, *Games with Rules*, involves children making and obeying rules within the play event. This form of play is most important for children to develop self-regulation skills. Figure 5.1 below presents a summary of these stages:

Vygotsky's perspective on play, connecting it to the social context in which a child is brought up, suggests that adults and older children should also be involved in the play to enable the younger children to model both the roles and the use of props. Vygotsky promoted the notion that play, as learning, should lead development, as opposed to the more traditional Piagetian idea that development leads learning or play.

A most important aspect of play is that children create an imaginary situation. Vygotsky described this as follows:

> *I think that in finding criteria for distinguishing a child's play activity from his other general forms of activity it must be accepted that in play a child creates an imaginary situation. This becomes possible based on the separation that occurs, in the preschool period, of the visual and meaning fields.* (Vygotsky, 2016, p. 6)

He continued to discuss the importance of the imaginary situation and stated that:

> *From the point of view of development, the fact of creating an imaginary situation can be regarded as a path to developing abstract thought. I think that the rule-making connected to this leads to the development of actions, on the basis of which the division between work and play becomes established – a division which is encountered as a fundamental fact at school age.* (Vygotsky, 2016, p. 20)

A child's understanding of, and ability to, distinguish effectively between the imaginary situation and the real context is important in the development of abstract thought. Vygotsky ends his discussion by arguing how this develops into an understanding of external reality and thought:

> *At school age play does not die away but permeates the attitude toward reality. It has its own inner continuation in school instruction and work (compulsory activity based on rules). All examinations of the essence of play have shown that in play a new relationship is created between the semantic field – that is, between situations in thought – and real situations.* (Vygotsky, 2016, p. 21)

Vygotsky suggested that in play, children are involved in an imaginary situation, with explicit roles and implicit rules. For instance, when playing 'families' they take on clearly understood roles and their actions are determined by those roles. This leads to a greater degree of self-regulation, the children's actions being determined by the rules of the game. When involved in play, children's concentration and application to the task are much greater than in academically directed activities contrived by the teacher. In the next section, how this combination of imaginary situation and rules underlies much of the work that scientists engage in is considered.

To summarise, play in early childhood science which facilitates imaginary situations and rules prepares the way for children to develop both abstract thinking and self-regulation, two of the most important features of scientific work (see Fig. 5.2) below.

### 5.3.1 Cultural Mediation in Play and Development During the Early Years

Whilst it is a common observation that children learn from adults and other children, it is less obvious how this happens. Vygotsky suggested that the child appropriates cultural tools and ways to use them; the child interacts with the environment via the mediation of cultural agents. The child is the subject, not the object of learning.

The main cultural tool, according to Vygotsky, is language, which can be thought of as a sign system. Vygotsky noted the importance of cultural mediation of these sign systems in humans, which does not appear to occur in animals. For instance, in the everyday activity of eating, animals of a particular species all eat in the same way, whereas in humans, the way a person eats strongly reflects the culture in which they were raised and there are many ways in which humans consume their food. Vygotsky maintains that cultural mediation is just as important in the consideration of how, and indeed what, children learn.

In terms of learning, the 'mediator', for example language, carries *meaning and sense*, as well as functioning as a tool, and must therefore be *interpreted* by the child (Zinchenko, 2007). Through this interpretation, the child contributes to the culture, and continues this contribution in many ways throughout their life. The use of language, ZPD, and imaginary play support a child's mediation with the culture as they develop in the contexts of their family, school, local culture, and global environment.

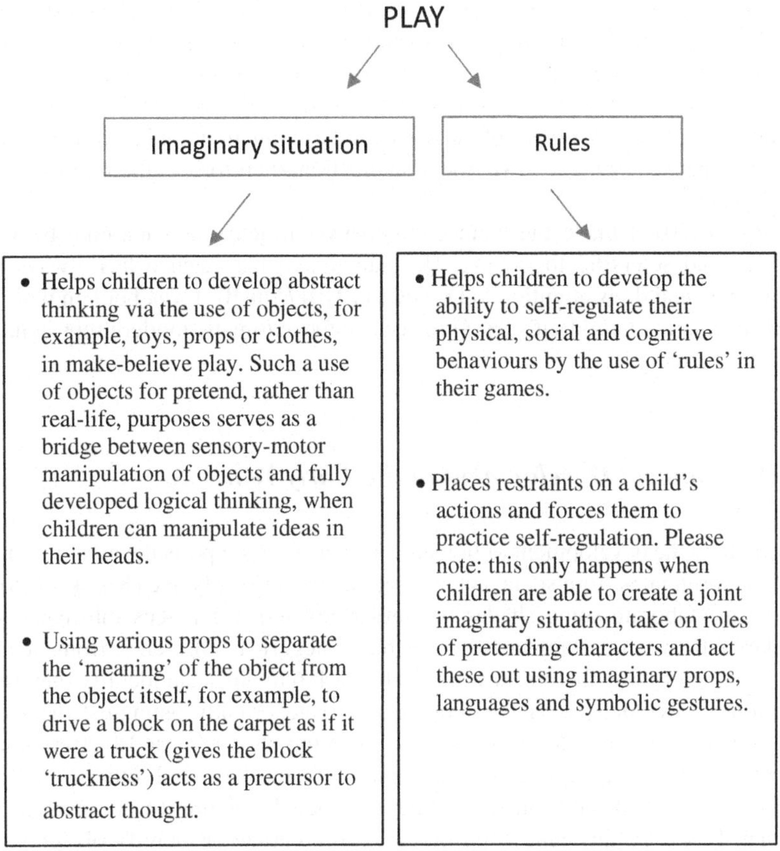


**Fig. 5.2** Ways in which imaginative play with rules is a precursor to academic scientific learning and practice

## 5.4 Scientific Concept Development Through Play

In preschool and primary school contexts, a Vygotskian perspective presupposes that teachers promote role-play and imaginary play in learning for children throughout the school to further the development of abstract, conceptual thought. There is less focus on individual play with objects and more on collective play, preferably involving older children who can model both roles and the use of props for the younger ones. The older children's learning develops strongly as they 'teach' the younger ones, just as teachers improve their learning as they describe concepts, such as photosynthesis, in ways that students can understand.

Vygotsky viewed development as a revolutionary or transformational process. He employed the idea of all processes being always present but being "on the stage" and leading at different periods to highlight the complexity of development (1983, p. 145). The process of development contains not just "evolutionary but also revolutionary changes, regression, gaps, zigzags, and conflicts" (Vygotsky, 1997, p. 221). The child, according to Vygotsky, will experience periods of rapid development, as well as tensions and struggles, as they interact with their environment. For Vygotsky, "psychological functions are given in the form of social relations which are the source of the origin of these functions and their development within humans" (1998, p. 473). Social interaction, therefore, is the root of cultural development.

Cowles (2017) made the connections between elementary learning by young children and scientific theorising. He maintained that science has always been child's play and drew attention to Dewey's (1910) *How We Think* book in which his short schematic of children's learning became the axiomatic modern representation of scientific thought.

### *5.4.1 Concept Development in the Early Years*

Research on the development of human consciousness supports the assumption that meaning originates from social categorisation at a very early age (Köcski, 1981). A child can elaborate some shades of similarities and differences into categorical similarity between certain factors and their categorical difference from others if they are one of these factors. A young baby of 4–6 months can identify themselves with some individuals and, at the same time, can categorically distinguish themselves from others – this is often referred to as 'making strange'. An 18–20-month-old child can distribute similar objects among themselves and others, and then distinguish each of them on the attribute of their belonging to one person or to another. However, the same child may not differentiate or identify objects on the attribute of their colours before the age of 3 (or even, according to some authors, age 4 or 5).

In terms of more abstract categories, such as length and volume, Piaget's classical experiments showed that children below the concrete operational stage (7–11 years old) cannot conserve length or volume. For example, regarding number, younger children faced with observing the experimenter setting out two rows of, say, coins of which one is tightly packed whilst the other shows the coins widely spread, when asked which row contains more, invariably respond that the one with coins spread out has more. On further questioning as to why, most children look perplexed and 'measure out' the rows with their hands, indicating that the row with spread out coins is longer. Similarly, most young (pre-operational) in Piaget's terms) children observing the same amount of water being poured into two different vessels, of which one is longer and thinner than the other, respond to the question of which vessel has more by saying that the taller vessel has more as the water level is higher. However, Doise et al. (1975) showed that pre-school children responded correctly to the questions asked about number and volume when *they carried out the tasks themselves*, revealing that *putting the child at the centre of the learning promotes meaning-making within concept formation*. Their findings represent the Vygotskian, cultural-historical underpinnings of concept development, in which children's social and physical environment are key to their meaning-making, and thus in concept formation.

Vygotsky and his co-workers proposed stages that are passed through in the formation of concepts by children. These stages comprise (a) '*random*' *grouping,* in which they categorise using their subjective experiences; (b) *thinking in complexes,* in which they start to abstract isolated features such as length and volume, together with the subjective ones; (c) *pseudoconcepts,* in which they apply correctly learned ideas in certain situations, such as tests, without an understanding or comprehension of links between different concepts; and (d) *true concepts*, which are bound by logical bonds within parts and between different concepts.

In the early years, teachers can maximise young children's opportunities for concept formation by ensuring that they can make sense of what is being taught in relation to their sociocultural contexts.

Ideas from Vygotsky, Piaget and their co-workers on concept formation were used for an analysis of scientific concept formation in 5–7-year-old children (Murphy, 1987). This study involved 280 pupils and 25 concepts, categorised as concrete or abstract, and derived from published word lists, and from infant science texts and worksheets. The concepts, presented as words, were used in a classroom game in which one child (volunteer) would be given a card with the concept word to hold and look at without it being seen by the other children. The volunteer child described the word (concept) without saying it aloud, until the class guessed it correctly. It was data recorded from the volunteer children that was recorded for analysis. Analysis comprised psychological factors (such as animism [Piaget, 1929, 1930]), the syntagmatic-paradigmatic shift, discrimination, and generalisation (Vygotsky, 1962); linguistic factors such as monosyllabic or restricted responses, and grammatical errors and analysis of responses with respect to age, gender and ability; and consideration of these in terms of biology, chemistry and physics learning.

One of the major findings indicated that both Vygotsky and Piaget seriously underestimated children's thought in terms of both coherence and systematic thinking. The high level of abstraction in this study evidenced different children describing the concept of amount as "degrees", weather as a "sort of condition", sound as "things you hear", transport as "a sort of vehicle", idea as "a plan" and a cat as "an animal with a tail and whiskers".

Key conclusions showed that the psychological factors of animism and syntagmatic-paradigmatic shift were not important features of concept formation. Discrimination was used as a strategy in developing concepts involving quantity, whereas generalisation was used in developing concepts representing stable or constant forms. The youngest children evidenced more separation of linguistic and concept acquisition, although the two processes became more intimately linked with the emergence of metalinguistic and metacognitive thought. In addition, concept development was observed as children creating 'rules' to categorise objects and/or events. Age was a more significant factor than both gender (no effect) and ability (no clear effect) in concept formation. Children as young as 5 were capable of coherent and systematic thought. The most important scientific concept for a young child to acquire as an introduction to scientific thinking is the concept of *change*.

Some scientific concepts are more difficult to acquire than others because of the need to make semantic and syntactical transformations of the concept before it can be considered in scientific terms. The author's study (Murphy, 1987) described above, led to the appreciation of children's scientific concept development as a dynamic, exciting process which provides excellent opportunities for teachers of young children to engage with science learning and teaching experiences with a diverse array of materials and the progressive consideration of certain key scientific concepts, most particularly that of change.

## 5.5 Early Years Science Learning in Vygotskian Schools

Golden Key schools are a type of 'experimental' school in Russia, in which the programmes are based on Vygotskian principles, together with those from Steiner's anthroposophy, Waldorf pedagogy, Swedish kindergartens and the English nurturing system (Kravtsov, 2010). Golden Key schools typically cater for 3–11-year-old children.

The rest of this section describes observations that took place on two summer school Vygotskian immersion courses which a group of international researchers (including the author) attended as part of an international group of scholars in a small town in Southern Russia, Belaya Kalitva. The group worked within the school every day and evening, experiencing classroom observation, lectures from Vygotsky scholars, and 'play' sessions after school, in which the teachers introduced the group to how they planned and carried out play-based learning and teaching in the school.

Curriculum and pedagogy in Golden Key schools is developed with the understanding that adults can and should have an important role in promoting play that supports development. A significant portion of the school day at a Golden Key school is centred on preparation for an *event* or *happening*. Preparation for an event includes an integrated approach to all areas of the curriculum around the theme or story of the event. Teachers help create an imaginary context for the children to engage in. The juxtaposition of the real and imaginary helps children to understand the world they live in and, as they reach school, to age develop academic skills and knowledge.

Most of the learning is centred on 'events'. The event witnessed by the authors at the start of the week was the arrival of a letter, delivered directly to the classroom (for more detail, see Murphy, 2012). This letter was from a wolf (teacher dressed as a wolf), desperately seeking help from the children to find his fairy tale – he had jumped out of the book and could not remember which tale he belonged to. A series of activities was based around finding 'clues' from which, eventually, the children could determine the correct tale for the wolf. Such activities, some of which were carried out in mixed-age groups involving older children helping younger ones, and other age-specific activities based on areas such as mathematics, science, verbal and spatial reasoning, geography, comprehension, history, and drama, etc., were enacted over the period of a week.

Children were fully engaged and many of the skills they demonstrated in enacting the traditional curricular requirements (such as mental arithmetic, compositional writing, and logical reasoning) were considered by the author to be rather advanced for the children's age groups. They seemed to be working "a head taller" than themselves (Vygotsky, 2016, p. 18) in the quest to find the wolf's tale. Children were engaged in experiments, hands-on experiences, readings, and discussions during the events. These activities are considered foundational to true scientific (academic) thought, especially when children are encouraged to *theorise* on their experience of observed phenomena, and our group observed all of these activities during the week. Some of the science activities we observed were as follows:

- Children, guided by the teacher, created a timeline that started with the beginning of life, and children oriented their science learning in time using this (see Fig. 5.3 below). The timeline spanned a whole wall in the classroom. For instance, they marked the times when various events occurred, including:
  - Dinosaurs roaming the Earth.
  - The discoveries of fire, the wheel, the solar system, electricity, and the moon landing.
- Using the timeline, they visualised life spans of large trees, humans and elephants to consider themselves in relation to older members of their family, ascendants and younger members.
- During the study of time the teachers created a 'time machine' and during their 'time travel' the children become aware of great scientific discoveries. They realised that there was a time before electricity was harnessed and explored a time with no cars and where horses and candles were used instead of cars and electric lights.
- The goal was to help children to experience, in imaginary play situations, life before these discoveries. The time machine also 'took' children into the future – allowing them to use their 'spacecraft' to travel to planets, solar systems, and

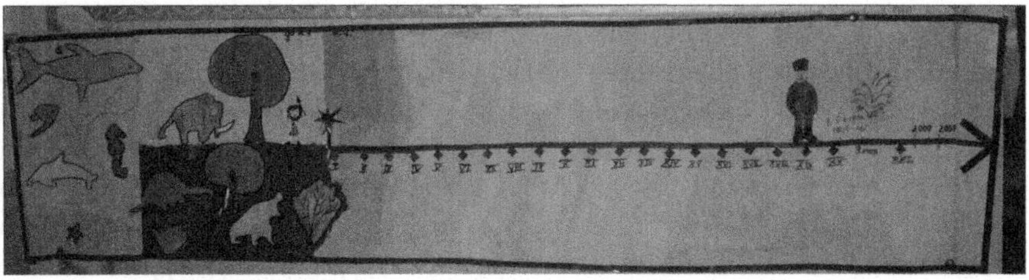

**Fig. 5.3** Timeline created by children to orient their science learning in time

galaxies. Through their imaginary travel, they investigated the cosmos and compared it to Earth.

- Later, young children were encouraged to design their own learning as a result of the teacher bringing in toy bear for children to play with it daily, take it on walks, 'feed' it, etc. Then, one day the bear was replaced with note saying it had gone north to visit its mother. The reaction from children was that they wanted to find the bear. With guidance from the teacher then, they asked for a map, prepare a 'trip' – How did they get there? What was climate like? What did they need? *Children here are the source of their learning* – we used the term 'spontaneous learning'.
- Solving the story of the bear (above), or in another case, the wolf (above, who jumped out of its fairy tale), included specific science problem-solving skills or the use of similar, for example, observation, communication, and analytical skills to decipher a letter which could help the wolf to find a location on a map. The development of these skills naturally supports future academic understandings and skills in all the subject areas. A fundamental assumption in this curriculum is that the experiences with space, both local and distant, through imagination and story, combine to provide an orientation of the world that is important in the child's future generalisation of theory and understanding of relationships of elements of the natural world.
- In other lessons, we observed children's use of materials in different ways depending on their age; for example, early exploration of materials is important for speech development, which manifests as the 3–4-year-olds playing individually and in groups with materials such as sand, mud, playdoh, water or toys.
- One example was children pouring syrup into a transparent container, then predicting where kitchen oil would settle when poured on top of the syrup. As expected, the oil formed a layer on top of the syrup. The third liquid to pour on top was water. Most children suggested that it would also form a layer on top. But the water settled in a layer between the syrup and the oil. They then repeated the experiment twice more, observing closely why the water moved underneath the oil. They discussed what happened and produced drawings and explanations to present to the class. Some groups observed that the oil had bubbles in it, and when the water was poured down the side, the bubbles 'lifted' the oil, which then moved underneath the oil. Such theorising is so helpful for children when they are introduced to abstract scientific concepts, such as density, when they are much older.

These experiences validate many pedagogical approaches which research has supported and open the door to investigating some interesting ways of thinking about providing science experiences for children from age 3 to 10 years. Such innovative and theoretically grounded approaches to create learning experiences in children's ZDP, to make use of imaginary experiences (play), and to help children connect with, and learn from, their culture and history, provide rich opportunities to bring new insights and ways of thinking about teaching and learning science in a variety of contexts. In the next section, I apply some of the Vygotskian ideas to digital learning in early years science.

## 5.6 Play and Digital Learning in the Early Years

Digital play has been described as: "the first qualitatively different form of play that has been introduced in at least several hundred years" and that "it merits an especially careful examination of its role in the lives of children" (Salonius-Pasternak & Gelfond, 2005, p. 6). The emerging phenomenon of digital play differs from that of children's spontaneous play, as the former largely depends on (and is often restricted by) the actual design of the software and hardware (Verenikina & Kervin, 2011). Another feature of digital play in the early years is that it seems to be used mostly for learning, as opposed to pleasure only. For example, studies of 4–5-year-old pre-schoolers demonstrated that educational games are used more frequently than recreational. This section explores children's digital play and how it relates to spontaneous play in terms of children's development and learning in the early years.

Cultural-historical theorists have drawn attention to the overarching role of play in child development and view it as the most significant, leading activity of the early childhood years (e.g., Bodrova & Leong, 2007; Vygotsky, 1967). Children acquire the foundations of self-reflection and abstract thinking, develop complex communication and meta-communication skills, learn to manage their emotions, and explore the roles and rules of functioning in adult society, whilst engaging in play. Imaginary play constitutes the basis for the child's awareness of the world around them and raises their cognition of reality to a more complex and generalised level. Vygotsky suggested that imaginary play marked the beginning of higher psychological functioning and abstract thought (Vygotsky, 1978). So, how does digital play measure up? Verenikina and Kervin (2011, p. 6) outlined dimensions of such play which set criteria for digital play to demonstrate for it to benefit children's development similarly to spontaneous, non-digital play, as:

- *dimension/s of pretend an action* and interaction in an imaginary situation
- *the use of substitutes*
- *spontaneous, self-initiated and self-regulated activity*
- *not goal-oriented*
- *relatively risk-free*
- *intrinsically motivated*
- *child in control.*

Verenikina and Kervin encouraged software designers to understand the richness of children's traditional play and use its developmental advantages in their products, such that parents and early childhood educators can access software which is suitable for the context in which it is being used. A feature of traditional play is that children often act out situations in real life according to how they relate to their own experience. An example is two four-year-old children, who were in the house when their grandmother had a cardiac arrest. Later, they 'played back' the situation using a scene from a TV programme they had seen, in which a patient's heart was resuscitated using electrical stimulation, mimicking the 'violent' actions required.

Research carried out by Verenikina and Kervin (2011) showed children taking ideas from digital games and creating physical representations during later periods of play. Often, collaboration with adults or older children is needed for young ones to navigate the software, so that they can ultimately be in control of the game. Children preferred digital games which allowed them to "engage their imagination and develop their own play that extended beyond the screen as digital play blended in the variety of children's other play contexts" (Verenikina & Kervin, 2011, p. 17). Other research by Verenikina et al. (2010) identified characteristics of digital gaming that supports higher order thinking and facilitates developmental play, including that it:

- *is intrinsically fun and not limited in scope to 'teaching' particular skills*
- *allows play for the sake of play – reaching goals is less important*
- *relates to daily life – sounds and objects from daily life and other things that the child can recognise; the actions of the characters and the rules of their behaviour are relevant to the context of children's real life*
- *can be incorporated into children's imaginative play*
- *is discovery-oriented*
- *allows children choices in the selection and timing of activities; allows for multi-functional use of the objects represented on the screen*
- *allows the manipulation of symbols and images on the computer screen*
- *provides the facility to engage collaboratively with the programme rather than exclusively in single player mode*
- *provides visible transformations on screen*
- *enables increasing complexity*
- *provides spoken directions (as children may not be old enough to read), or provides advice that children need assistance from more experienced players*
- *employs an uncluttered screen design with simple background, colouring, and graphics.*

### *5.6.1 How Do Children Learn to Use Digital Technologies?*

Another important aspect of learning and digital play in the early years is how children learn to use digital technologies. Considering Vygotsky's (1997) ideas regarding tool-mediated activity and Hutt's categorisation of children's epistemic (explorative) and ludic (imaginative) play, Edwards and Bird (2017) have developed a Digital Play Framework, which can be used by educators to observe and assess young children's digital activity. This framework sets out indicators of epistemic and ludic play. In epistemic play, children are the technology, and indicators include locating and testing different buttons, asking for help with navigation and sharing new knowledge with other children. Ludic play evidence mastery of the technologies, and Edwards and Bird's (2017) Digital Play Framework indicators include use of a device to record imaginative play, selecting a programme for imaginative play and creating such play deliberately for the use of the device. Thus, educators can map children's development and learning via the use of digital tools, as they observe them at play.

The most important influence of imaginative play on children's development is the evolvement of psychological representations and the separation of thought from concrete actions and real objects (Vygotsky, 1967). The make-believe situation of play creates an imaginative dimension in which the child uses symbols and signs to substitute for objects and acts. According to Vygotsky (1967, p. 12):

> *Thought is separated from objects because a piece of wood begins to be a doll and a stick becomes a horse. Action according to rules begins to be determined by ideas, not by objects. This is such a reversal of the child's relationship to the real, immediate, concrete situation that it is hard to evaluate its full significance.*

Thus, within an imaginary situation, children separate the literal meaning of the object or situation from its imagined or symbolic meaning, an early manifestation of the development of abstract thought, which is vital in science learning. Remote control and other toys which are increasingly lifelike, can constrain imaginative thinking somewhat. Content-free technological tools, which enable children to create their own imaginary situation, hold more potential for developing children's abstract thinking.

### *5.6.2 Preschoolers, the Internet, Science Learning and Cyber-Safety*

Research into early childhood science education shows that children's contextual experiences are critical to determining their understanding of a given phenomenon. For example, the concept of chemical change can be contextualised for children in terms of cooking food. Young children can be led to understand that food changes state when heat is applied, and it is cooked. Edwards et al. (2018) suggested that in science education, teachers now focus attention on children's contextualised experiences rather than assuming 'misconceptions' because their ideas are qualitatively different to those held by scientists. In terms of conceptual understanding of the internet, Edwards et al. (2018, p. 47) consider that:

> *...**bounded** social and technical concepts of the internet may be more a function of using adult concepts of the internet as a yardstick for children's understanding than they are a reality of what children actually think or understand about the internet.* (Emphasis in original)

In terms of Vygotsky's (1987) CHT, Edwards et al. (2018) argue that children's "internet cognition" (p. 48) may be more effectively understood according to their everyday concepts of the internet itself. Vygotsky proposed that scientific concept development occurs within the everyday/abstract dialectic, thus introducing children to a more informed idea of the internet as a complex computer network which provides information, entertainment, and communication facilities.

Internet experiences from both home and school will contribute to children's understanding and use of internet applications. Research on internet use with very young children in nursery/primary schools investigated the use of virtual learning environments (VLEs) for science experimentation (Beggs et al., 2009). A group of children in different schools were linked together online via Blackboard and were investigating 'sidedness' (Greenwood et al., 2007) by recording each child's dominant hand, foot, eye, ear, etc., in a series of activities including jumping, hopping, throwing, listening, and looking. The VLE enabled the children to observe pictorial representations (bar charts) of both the data of their own class and that of the whole sample simultaneously. This project also exhibited the less anticipated finding that teachers were more fully engaged in this type of learning than they had been from off-site CPD in digital learning.

Young children can conceptualise and think about using internet features for various purposes through their home and school experiences of internet use. If they are also educated in terms of the global networking of computers which comprise the internet, they can also be facilitated in learning about the dangers inherent in accessing and communicating data which can be unsafe in many ways.

## 5.7 Summary and Conclusion

This chapter has illustrated the huge impact of Vygotsky's work on changing the nature of teaching very young children. It foregrounds his extensive work on imagination and play and their importance for future learning and development.

Vygotsky also identified the very early age that children develop symbolic, abstract thought as the use neutral objects to represent an object or animal in play – examples are 'riding' as stick as a horse or creating a garage or school from a box.

Their use of trial and error and other forms of experimentation suggests that scientific method is innate and develops according to the cultural environment in which children are situated.

The chapter has also described early years learning in a Vygotsky school in Russia, and how some of that learning has been implemented in an Irish school, with transformative effect (Doherty, 2013). Young children's play and digital learning were also discussed in terms of the benefits and challenges. This chapter also highlighted the strong impact of early learning on future child development.

The next two chapters focus on children's experiences of science learning in primary and secondary education, and on the importance of Vygotsky's work there, mostly in relation to the ZPD and aspects of scientific concept development.

## References


Ackerman, D. (1999). *Deep play*. Random House.

Beggs, J., Murphy, C., & Kerr, K. (2009). Inspiring Einstein minds. *Primary Science, 109*, 17–21.

Bodrova, E., & Leong, D. (2007). Playing for academic skills. In *Children in Europe: Vygotsky edition* (pp. 10–11). Children in Scotland.

Bronfenbrenner, U., & Morris, P. (2006). The ecology of developmental processes. In W. Damon & R. M. Lerner (Eds.), *Handbook of child psychology volume 1: Theoretical models of human development* (6th ed., pp. 793–828). Wiley.

Cowles, H. (2017). Child's play. *Aeon Magazine*. Retrieved from https://aeon.co/essays/how-the-scientific-method-came-from-watching-children-play

Dewey, J. (1910). The analysis of a complete act of thought. In *How we think* (pp. 68–78). D.C. Heath.

Doherty, A. (2013). *Vygotsky and play: A critical exploration of theory into practice*. Unpublished doctoral thesis, Queen's University Belfast.

Doise, W., Mugny, G., & Perret-Clermont, A. N. (1975). Social interaction and the development of cognitive operations. *European Journal of Social Psychology, 5*(3), 367–383.

Edwards, S., & Bird, J. (2017). Observing and assessing young children's digital play in the early years: Using the digital play framework. *Journal of Early Childhood Research, 15*(2), 158–173.

Edwards, S., Mantilla, A., Henderson, M., Nolan, A., Skouteris, H., & Plowman, L. (2018). Teacher practices for building young children's concepts of the internet through play-based learning. *Educational Practice and Theory, 40*(1), 29–50. Retrieved from https://doi.org/10.7459/ept/40.1.03

Fleer, M. (2012). Imagination, emotions, and scientific thinking: What matters in the being and becoming of a teacher of elementary science? *Cultural Studies of Science Education, 7*(1), 31–39. Retrieved from https://doi.org/10.1007/s11422-011-9365-z

Greenwood, J. G., Greenwood, J. J., McCullagh, J. F., Beggs, J., & Murphy, C. (2007). A survey of sidedness in Northern Irish schoolchildren: The interaction of sex, age, and task. *Laterality, 12*(1), 1–18.

Hao, Y., & Fleer, M. (2016). Creating collective scientific consciousness: A cultural-historical study of early learning about Earth and space in the context of family imaginary play. *Asia-Pacific Journal of Research in Early Childhood Education, 10*(2), 93–124.

Hayes, N., O'Toole, L., & Halpenny, A. M. (2017). *Introducing Bronfenbrenner: A guide for practitioners and students in early years education*. Routledge.

Hedegaard, M. (2009). Children's development from a cultural-historical approach: Children's activity in everyday local settings as foundation for their development. *Mind, Culture, and Activity, 16*(1), 64–82.

Köcski, M. (1981). *Pozitsiia v sotsial'noi situatsii I psikhicheskoie razvitie rebionka* [Position in the social situation and child's mental development. A longitudinal study]. Unpublished academic thesis, Moscow State University.

Kravtsov, G. G. (2010). Cultural-historical/nonclassical basis of the 'golden key' school program. *Journal of Russian and East European Psychology, 48*(4), 61–75.

Kravtsova, E., & Maximov, A. (2014). Play in the non-classical psychology of L.S. Vygotsky. In L. Brooker, M. Blaise, & S. Edwards (Eds.), *The SAGE handbook of play and learning in early childhood* (pp. 21–31). SAGE.

Murphy, C. (1987). *Primary science: An analysis of concept formation in 5-to 7-year-old children*. Unpublished M.Ed. thesis, Queen's University Belfast.

Murphy, C. (2012). Vygotsky and primary science. In B. J. Fraser, K. G. T. Tobin, & C. J. McRobbie (Eds.), *Second international handbook of science education* (pp. 177–187). Springer.

Piaget, J. (1929). *The child's conception of the world*. Routledge & Kegan Paul.

Piaget, J. (1930). *The child's conception of physical causality*. Routledge & Kegan Paul.

Salonius-Pasternak, D., & Gelfond, H. (2005). The next level of research on electronic play: Potential benefits and contextual influences for children and adolescents. *Human Technology, 1*(1), 5–22.

Scarfe, N. V. (1962). Play is education. *Childhood Education, 39*(3), 117–121.

Verenikina, I., & Kervin, L. (2011). iPads, digital play and pre-schoolers. *He Kupu, 2*(5), 4–19.

Verenikina, I., Herrington, J., Peterson, R., & Mantei, J. (2010). Computers and play in early childhood: Affordances and limitations. *Journal of Interactive Learning Research, 21*(1), 139–159.

Vygotsky, L. S. (1962). *Thought and language*. MIT.

Vygotsky, L. S. (1967). Play and its role in the mental development of the child. *Soviet Psychology, 5*(3), 6–18.

Vygotsky, L. S. (1978). *Mind in society: The development of higher psychological processes* (M. Cole, V. John-Steiner, S. Scribner, & E. Souberman, Eds.). Harvard University Press.

Vygotsky, L. S. (1983). *Collected works volume 3: Problems in the development in mind.* Izdatelistvo Pedagogika.

Vygotsky, L.S. (1987). Problems of general psychology. In *The collected works of L.S. Vygotsky, vol. 1* (pp. 243–285) (R. W. Rieber & A. S. Carton, Eds.; N. Minick, Trans.). Plenum Press.

Vygotsky, L. S. (1997). The history of the development of higher mental functions. In *The collected works of L. S. Vygotsky, vol. 4* (R. W. Reiber, Ed.; M. J. Hall, Trans.). Plenum (original work written 1931).

Vygotsky, L. (1998). The problem of age. In *The collected works of L. S. Vygotsky, vol. 5* (pp. 187–205) (R. W. Reiber, Ed.; M. J. Hall, Trans.). Plenum.

Vygotsky, L. S. (2016). Play and its role in the mental development of the child. *International Research in Early Childhood Education, 7*(2), 3–25.

Zinchenko, V. (2007). Lev Vygotsky: From "silver age" to "red terror". In *Children in Europe: Vygotsky edition (5–7)*. Children in Scotland.

# Chapter 6
# Vygotsky and Primary School Science

## 6.1 Introduction

This chapter examines some of Vygotsky's ideas in relation to children's science learning at primary (elementary) level (aged between 4 and 12). The focus is on the ZPD and its potential to enhance science learning and engagement in primary schools. The chapter gives an overview of Vygotskian and Piagetian ideas about young children's learning. Both researchers maintained that children's minds work in different ways, using different means, from those of adults. Piaget claimed that researchers need to discover innate, internal laws that govern the child's mind, whereas Vygotsky highlighted the importance that culture plays in determining a child's development.

Vygotskian theory also suggests that young children need to be given opportunities to theorise on natural phenomena whilst at primary school, to develop scientific thinking skills; for example, enabling children to generate explanations of phenomena which are consistent with their observations. If given the appropriate guidance and resources, children's observation and reasoning skills can be developed to a relatively high level. This chapter demonstrates, with examples, how teachers can create ZPDs which give children opportunities to dialogue, present and evaluate their ideas and theories about scientific phenomena, thus enacting scientific method. Three Vygotskian principles relate most closely to science learning in primary school, and which are most useful for science teachers and teacher educators to consider in their own designs for science learning and teaching with young children. These are: the ZDP; the importance of play in developing abstract thought; and cultural mediation.



C. Murphy, *Vygotsky and Science Education*,
https://doi.org/10.1007/978-3-031-05244-6_6

## 6.2 The Zone of Proximal Development (ZPD)

There is currently much discussion and debate about what Vygotsky meant by the 'ZPD'. The simplistic definition of the ZPD found in many textbooks and other publications suggests it is the gap between what a child can achieve unaided and with help. This definition could be said to imply little more than the fact that teachers need to help children. The ZPD is far more complex. It has been claimed that the ZPD is probably one of the most used and least understood educational concepts (Palincsar, 1998). Some researchers (e.g., Van Oers, 2007) have discussed the complexity of the ZPD and showed how the concept was an evolving notion even during the short research life of Vygotsky, who used it initially as an index for intellectual potential and later as an educational concept focusing on the conditions needed to establish a ZPD.

So – perhaps the best way to conceptualise the ZPD is to think of it as all the actions and interactions that need to take place in moving from one stage of development to the next (Fig. 6.1).

And what might be inside the ZPD 'cloud'? Vygotsky (2016) proposed that for each stage, a predominant, or *leading activity*, is required, together with several other peripheral lines of development. In his work on child development, Vygotsky pointed to *play* as the leading activity for development at pre-school and that *learning* is the leading activity in school-aged children's development. The *leading activity* cannot have effect without peripheral lines of development, which include social, emotional, biological, and environmental factors and activities.

Alongside the leading activity and peripheral activities, there is a *set of conditions* which must be present for progression through the ZPD. These conditions are summarised below, as they apply in primary education.

(i) *Interaction of Real and Ideal Form*

The ideal form is vital for development. In other words, we need to know where we are going, or we might never get there. For example, a baby crawling may catch sight of a slightly older toddler, who reaches a desired object more quickly by walking or running. This is the 'ideal' to which the crawling baby conceptualises as their aim. Thus, in primary school, teachers can design activities which enable children to 'see' the next stage of development. An example is showing children a video of older children engaging in an activity which the younger ones will be doing. They 'imitate' the older ones, which helps them to progress.

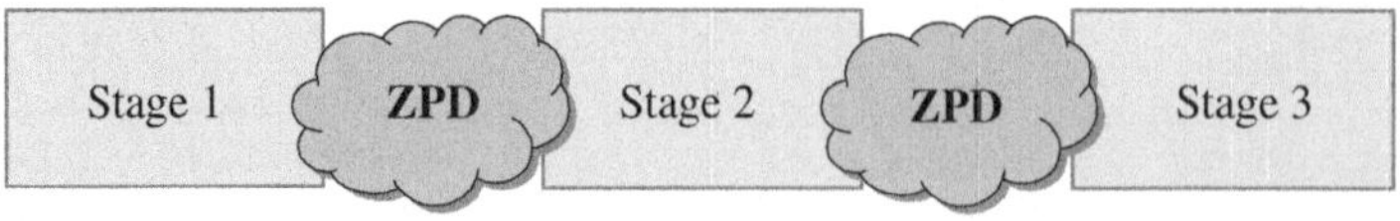


**Fig. 6.1** Conceptualised ZPD as all actions/interactions required to enable progress from stage to stage

(ii) *Buds of Development*

Buds of development represent the domain of learning/development that it accessible to the child, within the child's ZPD. For instance, it could be argued that understanding Schrödinger's mathematical formulation for studying quantum mechanical systems would be outside the domain of accessible functions of most primary school-aged children. So, it is outside the child's ZPD.

Vygotsky defined the ZPD as: "those functions which have not yet matured but are in the process of maturing… 'buds' or 'flowers' of development rather than 'fruits' of development" (Vygotsky, 1978, p. 86). He suggested that teaching/learning in the ZPD creates new levels of learning, or cognitive development, that would not have been reached otherwise.

The best learning occurs within the ZPD when the learner is at a stage, a bud or flower according to Vygotsky (1978), which is proximal (or close to), the next level of development. Such ZPDs can be created using cultural tools to promote learning readiness.

(iii) *Vygotskian Imitation*

Vygotsky's notion of 'imitation' is not exactly copying, more like *emulation* of an activity; the child is using imitation for the purpose of learning. Examples could be a child 'imitating' a teacher in knitting or crochet techniques, or indeed imitating chess moves effectively in learning to play chess. Each of these examples indicates that the child is 'copying' the teacher, but making their own product, or playing their own game. Eventually the child will perform such tasks without help, and with practice, they will reach the aim of a finished product or becoming a better chess player.

(iv) *Unity of Affect and Intellect*

The *unity* of affect and intellect in Vygotsky's ZPD, that emotion and learning are interdependent, foregrounds the importance of emotion in learning. Vygotsky stated that there is an: "intimate connection and dependency that exists between the development of the emotions and the development of other aspects of life" (Vygotsky, 1987, p. 332).

Around the age of 7, children recognises the difference between "feeling hunger and knowing that [they are] hungry" (Vygotsky, 1998, p. 291) both because they are learning to generalise the experience of hunger along with the word and because they are developing conscious awareness of themselves as experiencing hunger with similarities and differences from others who experience hunger.

Emotional engagement is required for a learner to maintain attention (Reid, 1788/1969). Teachers can work with children to provide contexts of learning which will ensure that the children will utilise voluntary attention to the activity. Rich learning environments, in which children are emotionally engaged, are required for effective learning of science in primary school.

(v) *Regression/Recursion*

Learning can be difficult and does not always assume a smooth upward trajectory. Regression is key to deep learning. Tharp and Gallimore (1988) proposed a four-stage model of the ZPD that addressed the development of any performance capacity based on the relationship between self-control and social control in an activity, which includes a 'recursive loop', in which learners revert to an earlier stage and progress through subsequent stages back to where they were – in effect they 're-learn' (see Fig. 3.1).

The core aspect of the ZPD described by Robbins (2003) is *personal transformation*, which is not always positive. In fact, she suggests that regression must occur for real growth and development. The element of regression and recursion allows teachers and children to acknowledge that learning is hard work. Another important aspect of the ZPD is a metaphor suggested by Zebroski (1994), who used the image of a tidal wave in relation to Vygotsky's work on development as a process that is progressive and regressive at the same time.

So, how is the ZPD characterised in the primary classroom, when children are learning science? Bert van Oers (2007, p. 15) pointed out that the ZPD "is *not* (emphasis added) a specific quality of the child, nor is it a specific quality of the educational setting or educators… it is… collaboratively produced in the interaction between the child and more knowledgeable others". The interactions required for ZPD creation involve both the material, as well as the social, learning environments. We need to create both social and material environments, aimed specifically towards children's learning.

## 6.3 Creating ZPDs to Enhance Science Learning in the Primary Classroom

In Chap. 5, play was a major focus for early years learning and development. Vygotsky argued that play is also a major feature in primary school education, particularly when combined with the creation of ZPDs.

> *At school age play does not die away but permeates the attitude toward reality. It has its own inner continuation in school instruction and work (compulsory activity based on rules). All examinations of the essence of play have shown that in play a new relationship is created between the semantic field – that is, between situations in thought – and real situations.* (Vygotsky, 2016, p. 21)

Vygotsky's description of the ZPD was that of maturing psychological functions that are required for the understanding of more abstract, scientific concepts (such functions are known as HPFs). The conditions required to create a ZPD to promote maturation of these functions is of prime importance to children's early development of scientific concepts. It is important that children should learn more about the *process* of science in primary schools, as a strong foundation for more conceptual learning at secondary level. Below is an example of teachers developing an activity in which children imitate the way scientists work.

(a) ***Children Imitating Ways Scientists Work***

An example of a bridging activity which is designed to aid children in developing their own theories about phenomena using close observation is described in part of my research on children's motivation to learn science (Murphy et al., 2013). Children (aged 7–8) were introduced to the phenomenon of miscibility via a teacher demonstration of pouring syrup, then oil and then water into a glass jar. They observed that the water formed a layer between the syrup and oil, a phenomenon which they might not have expected. They then repeated the experiment in groups a few times (to introduce them to scientific replication) to observe very closely and see if they could come up with an explanation, based on their observations, for water displacing the oil (see Fig. 6.2).

Following the experiment, children presented their theories to the rest of the class to explain the phenomenon, using diagrams, writing and, if appropriate, presentation tools, mimicking scientists presenting their experiments at conferences. This aspect of the activity expanded the ZPD created by the teacher to enable the children to work as scientists in this activity. An extract from one of the children's group presentations to explain the reason why water formed a layer between syrup and oil was:

> *...the cooking oil is at the top and the liquid ... there [were] bubbles in the cooking oil, and it is free, like, it can move around and then it, amm, lifted up and then the water went underneath it.* [8-year-old child]

This level of close observation and generating explanation consistent with the observations is rare, even at higher levels.

Recently, some of my secondary-level science student teachers repeated the above experiment. They were encouraged to present explanations based solely on observation, not inference (e.g., no scientific terms, such as 'density'). They found the task extremely challenging and were absorbed totally in the activity. Indeed, they commented that this approach to science learning and teaching was one to

**Fig. 6.2** Children engaged in scientific inquiry

which they had never been exposed, but from which they learned a lot about talking science with students of all ages.

Primary school teachers can be encouraged to promote activities which engender children's curiosity and close observation, as opposed to asking children to learn facts. Such an approach would require assessment which focused on scientific reasoning, which would provide an excellent foundation for post-primary/tertiary level science learning about conceptual frameworks which have been developed by scientists to explain phenomena.

In our work on children acting as scientists, we invited teachers of 8–9-year-olds and their pupils to 'create' ancient animals from a selection of fossils (Murphy et al., 2010). They were tasked to find out as much as they could using the internet and other resources during this activity, so that they would be carrying out this work in the same way that palaeontologists reconstruct animals and ecosystems from the past. Questions children asked during this activity indicated a strong interest in knowing more and more as they learned. Their awareness of how scientists worked in this field evidenced a contemporary view of the nature of science in which children described the work of scientists as systematic, but involving imagination and creativity. ZPDs were created again to encourage social learning and meaning-making. Most comments from children doing this work used the first-person plural 'we' in their responses when reflecting on these lessons. They expressed a view of science that made more sense to them. One child suggested in response to a question about scientists' work that: "Scientists are not always right, but they do their best" (Murphy et al., 2010).

These examples support the idea that primary school science needs to be relevant both in terms of children's everyday scientific experience and in terms of the world of science. Relevance to science in the world has received very little attention in discussions of the relevance of primary science, which tend to focus only on relevance to everyday life. How else then can young children be enabled to construct their world? How can teachers provide answers to their many 'why' questions? It seems, from this evidence, that activities which link school science to the work of scientists by creating ZPDs can help to provide this missing link.

(b) ***Children Acting 'Higher' than Themselves***

Other examples of activities of ZPD creation to promote dialogue and presentation came from giving children opportunities to express their ideas of how things might 'work' (Murphy et al., 2013). A 'black box' activity introduced by Hans Persson called 'The Bucket' was extended by teachers with a class of 6–7-year-old children, available on YouTube[1] (see footnote). Clear water is poured into a plastic bucket through a plastic funnel – the water then flows out of the bucket via a plastic tube (inserted into a hole on the side of the bucket) into a container. Red and then green-coloured water (food colouring) is poured into the bucket, and each time clear water flows out (see Fig. 6.3).

[1] https://www.youtube.com/watch?v=7tJkltfO4tc

**Fig. 6.3** The bucket! (Copied from Hans Persson – personal communication)

**Fig. 6.4** A child explaining how his car works

Some teachers extended the 'bucket' activity by starting with a sorting toys activity, followed by observation of the movement of a wind-up toy car. Children were invited to draw what they imagined was the inside of this car to illustrate what was inside the car to make it move. They then presented and explained their drawings to the class. The school principal overheard some children and created a ZPD for them to present at a higher level by offering his office for the presentations to a video camera. Each child sat on the principal's chair in his office and spoke to the camera (see Fig. 6.4). Their descriptions were recorded and transcribed. A typical one was:

> *My name's ... and I'm going to show you how this car works. The power of the pump goes into the batteries and makes more power in the batteries. And then it goes into the wheels. Then you push the button, and it goes zoom and fast. And then this here is the engine, and these are the wires that are connected on to the engine... (boy, aged 6)*

This description revealed the way that children were thinking and bringing their experiences into the science classroom. One child highlighted his concept of 'power' in describing how, and how fast, the car moved. Amongst others, descriptions and pictures focused on the central function of cogs in turning wheels, and on electricity.

Video footage evidenced children's engagement with the task and their clear enjoyment of being given the opportunity to express their ideas in words as well as pictures.

The teaching sequence continued with the children planning how they might build a car, using a selection of provided resources, such as cereal packets and plastic wheels. They drew their plans and then built a prototype, which was tested and rebuilt accordingly. The final cars were raced, and each child evaluated their own using two features they liked and one they wished they had included ('2 stars and a wish' – see Fig. 6.5 below). Finally, children examined all the cars and selected their favourite feature from one of the designs. The use of scientific language was 'scaffolded' for children via the provision of opportunities to discuss in groups, prepare and present findings to peers as well as teachers, and answer questions on their project work. ZPD creation provides 'bridges' for children to cross from primary science towards inquiry-based conceptual learning at secondary and tertiary levels.

(c) ***Primary Science Fairs: ZPD for Schools, Children, and Teachers***

One way of creating a ZPD for entire schools, teachers and children is to involve schools in a large-scale public event to showcase their science. The effort requires an intensive focus on a science project in each school, which requires the commitment of all. The outcomes in terms of children's and teachers' science skill development and learning show how this approach to learning motivates children to work at a much higher level than their norm. The fair is run by one of the world's longest-serving philanthropic organisations, the RDS (Royal Dublin Society) which was founded in 1731 to help Ireland thrive economically and culturally by driving the creation and cultivation of potential in people. Their investment in the RDS Primary Science Fair is part of this commitment.

The RDS Primary Science Fair provides a public forum for primary schools across Ireland to showcase their class STEM (science, technology, engineering, and mathematics) projects, and receive feedback from experts in the field. Whilst the

**Fig. 6.5** Child showing her '2 stars and a wish' assessment of her car

core focus is on children's skills development, the fair brings together teachers, parents, and students to showcase class projects, share experiences and learn from each other. Thousands of children take part in this fair every year in Ireland, north and south. Schools are encouraged to send in proposals of a science project, which they will carry out over a period of months and present at the fair. All proposals must submit their research question(s), possible hypotheses based on the children's knowledge, evidence they might collect (e.g., measurements, observation, building, coding, making) and how they will present it (e.g., charts and classifications), interpreting their findings and questions they might consider, such as whether the findings were expected, what they might have done differently, and how their work related to real life. All schools submitting satisfactory proposals are invited to the fair, on the condition that the whole class must attend, not a small group.

Each school is provided with a stand, in which their school details and project title are provided. Four or five students from the class 'man' the stand for 30 min, whilst the rest of the class move to other stands and take notes on other projects, and attend 'rocket science' exhibits provided by local organisations and industries. Then the next group takes over, until all class members have carried out this role. Judges, journalists, parents, academics, and members of the public attend the stand and engage the children in discussions of their work. The experience as a judge is that the students are so enthusiastic, proud of their work, anxious to communicate and exhibit a higher level of verbal and written skills in terms of science than is the norm for children of their age. This work creates a ZPD for children to 'perform' their science, discuss it with experts in the field, and make plans for how they might develop their projects when they get back (see Fig. 6.6). Children also question the 'experts' about their projects. Other researchers have focused on questions learners ask. Archer et al. (2015) concluded that children's questions are key to revealing

**Fig. 6.6** A group of children discussing their project on making batteries from fruit and vegetables

much about their prior conceptions, interests, motivations, and development, and could be used to a greater extent in science learning.

The RDS Fair aims to support primary school teachers to develop their skills and to mentor other teachers in integrating STEM within their classwork. Independent evaluation of the fair has shown that 97% of participants reported that their science skills improved, and 80% noted an improvement in their mathematics.

(d) ***STEAM-in-a-Box (SIAB): Linking Industry and Primary Science Teaching***

'STEAM-in-a-Box' (SIAB) is a project run in Ireland by STEAM Education Ltd.,[2] which facilitates primary children's active involvement in a holistic experience of science via 10- or 25-week programmes co-taught in school by scientists and their own teachers. I am one of the five members of the Advisory Board of STEAM-Ed Ltd. My key role is to develop and run the induction sessions to groups of primary school teachers and STEAM experts, working together and co-planning of lessons. They also try out the classroom activities, so both teacher and STEAM expert will work together as coteachers, in the school classrooms (see Murphy, 2016, which is a small handbook on coteaching).

The term 'STEAM' represents the subjects of Science, Technology, Engineering, Arts and Mathematics. At the time of writing, SIAB has introduced programmes in all of the STEM areas, with Arts following. The pedagogy of SIAB is based firmly on Vygotsky's ZPD and on social constructivism, in which the latter introduces scientists and artists into primary school classrooms to work with the teacher in providing exciting, engaging, and high-level science lessons, aimed at children between ages 8 and 13 years. SIAB is designed to support, and not to replace, the primary/early secondary science curricula.

SIAB aims to inspire young children with a truly exciting experience of science at home and school. Research on coteaching science in primary schools shows that extraordinary results can be obtained through external specialists working closely with the usual classroom teacher (e.g., Murphy et al., 2010). SIAB adopted a coteaching model whereby a scientist (typically a PhD science student or a scientist working in industry) plans, teaches and evaluates science lessons with the schoolteacher. Scientists and teachers attend induction workshops at a venue away from school, in which they are introduced to coteaching, and they practice the STEAM activities with the resources that they will also use in the classroom. They also co-prepare the first couple of lesson plans.

Following induction, the scientist arrives at school once a week with a box of scientific materials, supported with engaging PowerPoint presentations and videos, and for an hour the coteachers engage the children in 'rocket science'. The programme content is designed to enhance the current primary school curriculum, while complementing and leading into the new Junior Cycle science programme, taught during the first 3 years of secondary education in Ireland.

[2] www.steam-ed.ie

The children receive a science journal at the beginning of the programme and are encouraged to follow up questions that arise in the classroom at home, do their own 'research' and attempt to find the answers themselves. Bringing science home with them is one way of extending their ZPD by encouraging reflection on their learning, collaboration with friends and family and linking science between home and school. The focus is on children's experience, and assessment is formative, via quizzes and games.[3]

Evidence from evaluating the first 3 years of SIAB shows that most (75%) of children reported that they were inspired by being taught high-level science, even though admitting that some of it was very difficult for them. In addition, children from schools in which the coteaching between scientist and teacher was more fully embedded were more positive about the experience. All of the school principals, schoolteachers and coteaching scientists responded that they felt that children were engaged fully in and really enjoyed the lessons; they also all responded that they would like to repeat the experience of coteaching SIAB. Many schools have now run the programme several times.

The potential for exciting and empowering children in the world of science at a young age could pave the way towards a future human culture in which there is a far higher level of scientific literacy than the present. SIAB, provides a programme of high-level science, co-taught by scientists and primary teachers incorporates nature of science both as an introduction and to support children throughout in their quest for scientific knowledge and understanding. The teacher-scientist partnership empowers children to appreciate science as it is in a way that they can begin to understand. There is no pressure to assess their knowledge of scientific facts; instead, the programme seeks to inspire, excite, and challenge the children. Future evaluation of this programme will seek to determine whether such early intervention impacts on children as they get older, in terms of subject choice. The evidence so far suggests that engaging primary school children with science in this Vygotsky-based manner encourages them to see and experience science as the dynamic, relevant, challenging, and systematic process of exploration that it is. Children become more aware of the scientific process and how scientists work, as they are introduced to the wonders of science in SIAB in an interesting and playful classroom environment.

The next section deals with Vygotsky's ideas on play as learning in primary schools.

[3] For more details of the SIAB programme, see Murphy et al. (2016).

## 6.4 Play in Primary School Science

There is much debate about play in primary science, as to whether the focus should be on teaching academic skills or engaging young children in make-believe play as a developmental activity. Bodrova and Leong (2007) suggest that there is a false dichotomy between play and academic skills when considered from a Vygotskian perspective. Vygotsky maintained that creating an imaginary situation in play is a means by which a child can develop abstract thought. Vygotsky (1933) discussed the importance of imagination in play and argued that it is the defining criteria of play, distinguishing it form other activities: "in play a child creates an imaginary situation" (Vygotsky, 1978, pp. 93–94). He further described an imaginative activity as creating a dual effect:

> *For example, the child weeps in play as a patient, but revels as a player. In play the child renounces his immediate impulse, coordinating every act of his behaviour with the rules of the game.* (Vygotsky, 2016, p. 15)

Vygotsky (1978) stated that "play creates a ZDP of the child. In play a child always behaves beyond his average age, above his daily behaviour; in play it is as though he were a head taller than himself" (p. 102).

The best kind of play to develop abstract thought involves children in using unstructured and multifunctional props, as opposed to those that are realistic. The former type of props strongly promotes language development to describe their use (e.g., a cardboard box can serve first as a shop, then as a school, then as home). Vygotsky proposed that this repeated naming and renaming in play helps children to master the symbolic nature of words, which leads to the realisation of the relationship between words and objects and then of knowledge and the way in which knowledge operates.

This type of play is not often seen in the classroom in school – many 3- to 5-year-old children are playing like toddlers, just manipulating objects, and not engaging significantly with other children.

### *6.4.1 Cultural Mediation in Play*

Whilst it is a common observation that children learn from adults and other children, it is less obvious how this happens. Vygotsky suggested that the child appropriates cultural tools and ways to use them; the child interacts with the environment via the mediation of cultural agents. The child is the subject, not the object of learning.

The main cultural tool, according to Vygotsky, is language, which can be thought of as a sign system. For learning to take place, language first needs to be internalised by the child. Vygotsky noted the importance of cultural mediation of these sign systems in humans, which does not occur in animals (see Fig. 6.7). For instance, in the everyday activity of eating, animals of a particular species all eat in the same

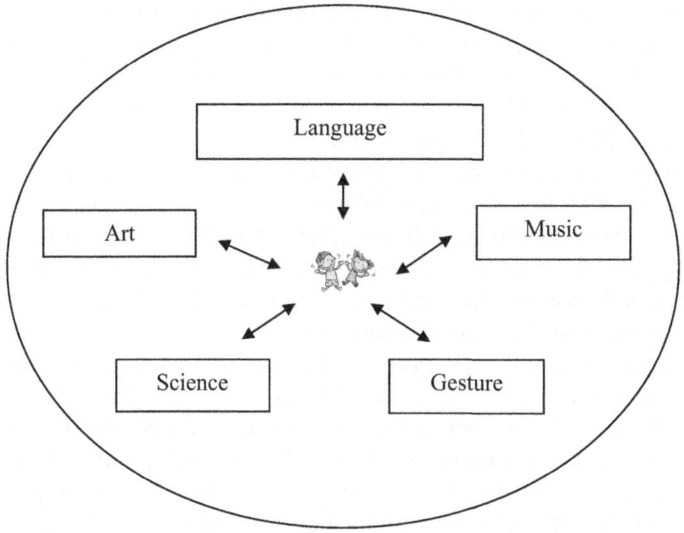


**Fig. 6.7** Examples of a sign systems used by a child to interact with the external world

way whereas, in humans, the way in which a person eats strongly reflects the culture in which they were raised and there are many ways in which humans consume their food. Vygotsky argues that cultural mediation is just as important in the consideration of how, and indeed what, children learn.

In terms of learning, it must be remembered that the 'mediator', such as language, carries *meaning and sense*, as well as functioning as a tool, and therefore must be *interpreted* by the child (Zinchenko, 2007). Therefore, the child contributes to the culture, and continues this contribution in many ways throughout their life. Children's learning by way of cultural mediation can be summed up as follows:

$$\textbf{Child} \xrightarrow[\text{meditated by cultural agent(s)}]{\text{interacts with environment}} \textbf{HPFs}$$

Vygotsky's concept of education comprises two fundamental forms of mediation: mediation via cultural concepts and mediation via social interaction, which can be considered separately, but are inseparable. It is through such mediation, according to Vygotsky:

> *That which the child turns out to be able to do with the help of an adult points us toward the zone of the child's proximal development. This means that with the help of this method, we can take stock not only of today's completed process of development, not only the cycles that are already concluded and done, not only the processes of maturation that are completed; we can also take stock of processes that are now in the state of coming into being, that are only ripening, or only developing.* (Vygotsky, 1956: 447–448, as cited in Wertsch, 1985, p. 68)

To aim the mediation at those abilities which are in the process of ripening, teachers can assess children's learning before and during, as well as after, each learning sequence. An emphasis on different modes of formative assessment, or assessment for learning (AfL) provides a basis upon which this may be achieved (see Black & Wiliam, 1998; Hardman & Teschmacher, 2019).

In primary science learning, the Vygotskian interpretation allows for the sharing of ideas about phenomena between children and their peers and teachers, which is essential for the exposure of different levels of understanding to be addressed. Vygotsky contended that higher psychological (cognitive) functions originate from the interaction between people, but there is also a need to *teach* decontextualised concepts to enable facilitation of cognitive growth.

Teaching decontextualised concepts can enable students to create and enliven a cognitive framework in which they can contextualise and abstract their experiences. For example, the fact that a person boils water in a kettle and observes steam coming out for years, does not necessarily (and only very rarely) lead to them discovering the concept of evaporation. Only when they are taught about evaporation and encouraged to link this learning with the kettle experience can most people make sense of the decontextualised concept of evaporation, and to situate other experiences, such as the drying up of puddles, within the initial framework of evaporation and then in the broader conceptual framework of the water cycle.

The next section presents a description of learning and teaching science in a Vygotskian primary school (ages 3–11) in Russia. Chapter 5 focused on activities suitable to younger children (3–7 years) in the same school. In this chapter, the focus is on science activities for children aged 8–11 years.

## 6.5 A Whole-School Approach to Vygotskian Science Learning and Teaching in Primary School

### *6.5.1 Summary of the Golden Key Programme*

The Golden Key programme developed from consideration of the work of several programmes and scholars, including Swedish kindergartens, the English nurturing system, Rudolf Steiner's anthroposophy and the Waldorf pedagogy, in combination with the seminal work on child development and learning carried out by Vygotsky, his co-workers and his followers. The founders of the Golden Key Schools (during the early 1990s) were themselves students of the actual followers of Vygotsky. Elena Kravtsova's teacher was A. V. Zaporozhets, and Gennadi Kravtsov was a student of Daniil El'konin. Elena is Vygotsky's granddaughter, who died in April, 2020; she will be very sadly missed.

The use of language, ZPD, and imaginary play support a child's mediation with the culture as they develop in the context of their family, school, local culture, and global environment. The Golden Key curriculum and pedagogy support this mediation through the deliberate creation of opportunities for children to actively engage with culturally significant events. Children may explore fairy tales, help an imaginary hero return home from another country, participate in national celebrations, or perform traditional dances. The juxtaposition of the real and imaginary helps children to understand the world they live in and as they reach school age develop academic skills and knowledge. The foundation of Golden Key schools is based on five key Vygostkian principles, summarised below:

1. ***Mixed Age Learning and Teaching***

   Golden Key schools adopt mixed age, as well as single age-group teaching, which allows the older children to reflect their own learning by helping younger children, who, in turn, benefit from learning from people closer to their own age and stage. The Golden Key approach justifies mixed-age classes by the relatively high achievement of children in their schools, when compared with the more traditional schools in Russia, which is partly due to older children teaching younger children. The older children reflect their own learning by helping younger ones and the younger children benefit from being taught by people closer to their developmental stage.

2. ***Family Principles – Closer Home-School Links***

   The schools are organised around family principles, including active parental involvement. Children are arranged into 'families', instead of into classes according to their age. As a result, each 'school family' has 15–25 children between the ages of 3 and 11. There is an atmosphere of an extended family where the children have their own 'home' instead of a classroom (Doherty, 2013). Golden Key schools appreciate that the teaching process is not limited to the classroom. They see parents as partners in education. Benefits of linking the home and the school to children include better levels of attendance at school, healthier attitudes to school, and better mental health. Benefits to teachers are better parent-teacher relations, enhanced teacher morale, and a more friendly school climate. Parents also benefit from increased satisfaction in their child's learning and improved confidence levels.

3. ***Paired Pedagogy***

   Paired pedagogy comprises two teachers taking some classes together. Typically, one acts as a traditional teacher, whilst the second acts in an 'under' role by, for example, asking naïve questions which are answered easily by the children. Sometimes this teacher has not been present earlier and is asking children to explain (and therefore reflect on) their learning; other times they take on the role of a 'fool', acting silly and requiring multiple corrections to their poor attempts to keep up. Murphy et al. (2009) carried out similar work using puppets, as opposed to adults, in the same role as the 'under' teacher.

4. ***Lessons Centred around 'Events'***

   Lessons are centred around 'events' which are highly meaningful to the children and touch their emotions. Subsequently, each lesson follows a 'plot' relating directly to the event. The 'event' we witnessed (as described briefly in the Chap. 5, was the arrival of a letter, delivered directly to the classroom. This letter was from a wolf, desperately seeking help from the children to find his fairy tale – he had jumped out of the book and could not remember to which tale he belonged). A series of activities (prepared by the teachers as a whole week's learning for all children) was based around finding 'clues' from which, eventually, the children could determine the correct tale for the wolf. Such activities, some of which were carried out in mixed age groups involving older children helping younger ones, and other age-specific activities based on mathematics, science, verbal and spatial reasoning, geography, comprehension, history, drama, etc., were enacted over the period of a week. Children were fully engaged and many of the skills they demonstrated in enacting the traditional curricular requirements (such as mental arithmetic, compositional writing, and logical reasoning) were very advanced for the children's age groups when compared with primary school learning elsewhere. They seemed to be working "a head taller than themselves" (Vygotsky, 1978, p. 102) in the quest to find the wolf's tale.

5. ***Unity of Learning and Development***

A key principle of Golden Key schools is the interaction and interdependence of education and development, so that single-age 'lessons' for younger children are not structured in the same way as those for older children. Learning takes place within their zones of proximal development (ZPDs).

## 6.6 Science Learning in Golden Key Schools

Children's learning takes place within the context of four major concepts, which define all human activity. All human activity takes place in a certain *space* (or place), at a certain *time*, uses *substances* (materials) and involves conscious human *reflection.*

All learning for children is oriented within this framework of concepts, and each year there is a focus on one, although all are addressed simultaneously to some extent. Experiments, hands-on experiences, readings, and discussions about science during the age of 3–12 are considered foundational to true scientific thought, especially when children are encouraged to *theorise on their experience of observed phenomena.* Below are some examples of science learning at primary school level in a Golden Key school:

### *6.6.1 Exploration of Place: On and Beyond Earth*

Teachers support children's ongoing exploration of space by creating imaginary journeys connected to the event that serves as the core of the lesson. They may plan, for example, a journey to rescue a hero from another country. These multi-age imaginary expeditions provoke many opportunities for children to engage in science learning in a wide ZPD. The teachers provide a context for the older and younger children to explore life, Earth, and physical SCs.

Teachers set up imaginary interactions with science phenomena during the children's 'travel'. As they go on their 'journey', they may look, for example, at which side of the rocks the moss grows or where the sun is – developing a connection with the moss and the sun. The older children may discuss these connections with the younger children. Children are introduced to, for example, the three states of matter by 'encountering' water as steam, water, and ice or snow in their adventure. A study of plants in their rooms may expand to learning about plants found in different parts of the world. Children may talk about which plants only grow in the north or only in the south, in sunlight or in shade, or soil characteristics, etc.

Every classroom had a set of large wall maps, superimposed on each other, so that children can orientate in the 'place' of all their learning; behind a map of the town was one of the provinces, behind this a map of Russia, then Europe and so on until the maps at the back were of the cosmos (See Fig. 6.8).

**Fig. 6.8** Maps to help children orient their learning in regard to 'place'

Not all the science explorations are clearly distinguished from other investigations of the children's world during their real and imaginary interactions with space. However, throughout the inquiry about the world in which they live, the development of their process skills also crosses 'academic' lines. Solving the wolf's dilemma may include specific science problem solving skills or the use of similar, for example, observation, communication, and analytical skills to decipher a letter as to help the wolf find a location on a map. The development of these skills naturally supports future academic understandings and skills in all the content areas including science.

The role of play cannot be overstated. A fundamental assumption in this curriculum is that the experiences with place, both local and distant, through imagination and story, combine to provide an orientation of the world that is important in the child's future generalisation of theory and understanding of relationships of elements of the natural world.

Children are also oriented within the concepts of time, materials, and reflection. Each classroom also had a timeline on a wall, created by the children, described in Chap. 5. Each timeline starts with the beginning of life, and children can orient their science learning in time using this. For instance, they can mark the times when dinosaurs roamed the Earth, the discoveries of fire, the wheel, the solar system, electricity or the moon landings. They use the timeline to visualise life spans of large trees, humans, and elephants and to consider themselves in relation to older members of their family, ascendants, and younger members.

### *6.6.2 Time Machine*

During the study of time the school created a 'time machine' and during their 'time travel' children become aware of great scientific discoveries. They came to realise there was a time before electricity was harnessed and explore a time with no cars and where horses and candles were used instead of cars and electric lights. The goal is to help children to experience, in imaginary play situations, life before these discoveries. The time machine also 'takes' children to the future, allowing them to use their spacecraft to travel to planets, solar systems, and galaxies.

Through their imaginary travel, they investigate the cosmos compare it to Earth. For example, children may compare the pressure, temperature, and length of day on Venus to Earth. The placement of the present day in terms of their cultural-historical context is viewed as important to allow and facilitate the children's mediation with their world and, in turn, promote development. As with the children's interaction with space, the imaginary and real interaction with time by a multi-age group and with the support of teachers provokes development and foundational (both real and imaginary) encounters with SCs.

### 6.6.3 Materials

In relation to the 'substance' concept, children use materials in different ways depending on their age. Early exploration of materials is important for speech development. As children get older, they focus on manipulating a wide variety of materials and theorising on these experiences to arrive at logical explanations of phenomena. Vygotsky maintained that children at primary level need to be encouraged in such activities for science learning, which are vital for the later development of conceptual thought within the concepts constructed by generations of scientists.

This world of science has its own 'culture' based on specific scientific tools such as signs and symbols, into which children will be encultured mainly at a later stage of their development (post-12) when they are taught by scientists or by teachers who have a good knowledge of science.

The early theorising about children's observations of phenomena is also how children become oriented within a framework of 'reflection.' Children are invited to present their ideas to other children and to their teachers, and to listen to and incorporate other ideas into their own reflections.

### 6.6.4 Reflection

The scientific methods and concepts switch from becoming the focus of science to the method of understanding and comparing and understanding different cultures. All the children have partially formed scientific understandings – emerging concepts, which become more complex as they get older. The science study becomes more focused as an academic subject.

An experiment is typically conducted with the whole (multi-aged) group at a Golden Key School. Older children also discuss experiments in their separate class periods. Then they come back to the whole group, and with the teachers' help, discuss the experiment with the younger children. This activity provides an enhanced ZPD for students.

The four concepts/themes (space, time, substance, and reflection) support the experiences and development of the children through intentional play experiences, learning leading development in the ZPD, and cultural mediation. A significant idea is that the development of these four themes is for the group rather than for individual children at a certain age. Holzman (1997) discusses the four themes at the Golden Key Schools. The organizing principle (philosophical structure) of this alternative educational model is the ongoing development of the group as *it creates itself as a group* that is developing and changing. This focus on the group is what – in Vygotskian fashion – allows for the "good learning" that the Golden Key documents for all its children.

The value of imaginary experiences and fully implementing the ZPD with "learning leading development" are not intuitive ideas. They, combined with the understanding that development is also a process of cultural mediation, rather than primarily a biological process, leads to schools, classrooms, and/or science classes that are very different from the typical. At the same time, the Golden Key approach validates many pedagogical approaches which research has supported and opens the door to investigate some interesting ways of thinking about providing science experiences for children from age 3 to 12.

These innovative and theoretically grounded approaches are used to create learning experiences in children's ZDP which make use of imaginary experiences (play). The aim is to help children connect with and learn from their culture and history, provide rich opportunities to bring new insights and ways of thinking about science teaching and learning in a variety of contexts.

In primary science, a Vygotskian perspective presupposes that teacher promote role-plays and imaginary play in science learning for children throughout the primary school in order to further the development of abstract, conceptual thought. There would be a lot less focus on individual play with objects and more on collective play, preferably involving older children who can model both roles and the use of props for the younger ones. The placement of the present day in terms of children's cultural-historical context is important to allow and facilitate children's mediation with their world and, in turn, to promote development. The imaginary and real interaction with time, under the guidance of teachers, promotes development and foundational (both real and imaginary) encounters with many SCs.

## 6.7 Summary and Conclusion

According to Vygotsky, learning *leads* development; so, do not wait until children are 'old' enough to learn. Leif Strandberg (2007) contends that teachers need to promote activities that:

- Develop interactions between children and between adults and children.
- Give children access to tools and words.
- Change around the learning environment to suit different activities.
- Involve children as creative co-workers.

Such methods liberate adults and children from a retrospective, diagnostic and resigned pedagogy, and enable a more forward-looking perspective on learning, with 'performing' as opposed to explaining. These methods also provide a sense of hopefulness for what comes next.

In primary science activities, teachers might consider expanding their use of curricular activities to include:

- Think, pair, share.
- Peer learning.
- Using mediational artefacts.
- Having a science term of the day (or week).

- Adapting the learning environment
- Using role-play and stories to promote Vygotsky-type imaginary play
- Extending 'play' activities to older children to aid abstract concept formation.

In summary, a Vygotskian approach to primary science highlights the importance of ensuring that practical activities are contextualised within a conceptual framework. Children are encouraged to discuss their developing understanding with peers and teachers, and time is allowed for experiences that foster the development of scientific concepts.

Role-play and collaborative, imaginative play with children of different age groups would be encouraged throughout the primary school to facilitate the development of abstract thought. Teachers may mediate pupils' learning by addressing social and cultural influences in their provision of appropriate educational tools and monitor children's progress as they attempt to identify and teach within their zones of proximal development. Teachers would use formal instruction *alongside* hands-on practical activities that are relevant to their experience and interests to enable children constantly to switch between everyday and scientific concepts until they have been adjudged to have achieved an appropriate understanding.

Changing the teaching/learning approach has required a level of theoretical synthesis between some of Piaget's ideas, which dominated much of the enactment of science teaching, and with the more operational aspects of Vygotskian theory. Social constructivism, partly based on some Vygotskian concepts, is mooted to be the most useful way of learning and teaching science in the early twenty-first century.

## References

Archer, L., Dawson, E., DeWitt, J., Seakins, A., & Wong, B. (2015). "Science capital": A conceptual, methodological, and empirical argument for extending bourdieusian notions of capital beyond the arts. *Journal of Research in Science Teaching, 52*(7), 922–948. Retrieved from https://doi.org/10.1002/tea.21227

Black, P., & Wiliam, D. (1998). Assessment and classroom learning. *Assessment in Education: Principles, Policy & Practice, 5*, 7–74.

Bodrova, E., & Leong, D. (2007). Playing for academic skills. *Children in Europe*, 10–11.

Chaiklin, S. (2003). The zone of proximal development in Vygotsky's analysis of learning and instruction. In A. Kozulin, B. Gindis, V. Ageyev, & S. Miller (Eds.), *Vygotsky's educational theory in cultural context* (pp. 39–64). Cambridge University Press.

Doherty, A. (2013). *Vygotsky and play: A critical exploration of theory into practice*. Unpublished doctoral thesis, Queen's University Belfast.

Hardman, J., & Teschmacher, N. (2019). Vygotsky's developmental pedagogy recontextualised as Hedegaard's double-move: Science teaching in grades 1 and 2 in a disadvantaged school in South Africa. In A. Edwards, M. Fleer, & L. Bottcher (Eds.), *Cultural-historical approaches to studying learning and development* (pp. 135–150). Springer.

Holzman, L. (1997). *Schools for growth: Radical alternatives to current educational models.* Lawrence Erlbaum Associates, Inc.

Jensen, E., & Wright, D. (2015). Critical response to Archer *et al.* (2015). Science capital: A conceptual, methodological, and empirical argument for extending bourdieusian notions of capital beyond the arts. *Science Education, 99*(6), 1143–1146.

Martin, S. N., & Scantlebury, K. (2009). More than a conversation: Using cogenerative dialogues in the professional development of high school chemistry teachers. *Educational Assessment, Evaluation and Accountability, 21*(2), 119–136.

Murphy, C. (2016). *Coteaching in teacher education*. Critical Publishing Ltd.

Murphy, C., Beggs, J., Kerr, K., & Persson, H. (2009, January). *Creative science teaching and assessment*. Paper presented at Association of Science Education, Reading.

Murphy, C., Murphy, C., & Kilfeather, P. (2010). Children making sense of science. *Research in Science Education, 41*(2), 283–298.

Murphy, C., Bianchi, L., McCullagh, J., & Kerr, K. (2013). Scaling up higher order thinking skills and personal capabilities in primary science: Theory-into-policy-into-practice. *Thinking Skills and Creativity, 10*, 173–178. Retrieved from https://doi.org/10.1016/j.tsc.2013.06.005

Palincsar, A. S. (1998). Social constructivist perspectives on teaching and learning. *Annual Review of Psychology, 49*, 345–375.

Reid, T. (1788/1969). *Essays on the active powers of the human mind*. MIT.

Robbins, D. (2003). Vygotsky's zone of proximal development, imitation and mimesis: Thoughts regarding L2 and psycholinguistics. *Kul'turnoistoricheskii Podkhod i Problemy Tvorchestva*, 52–57.

Strandberg, L. (2007). Vygotsky, a practical friend. *Children in Europe*, 16–18.

Tharp, R. G., & Gallimore, R. (1988). *Rousing minds into life: Teaching, learning, and schooling in social context*. Cambridge University Press.

van Oers, B. (2007). In the zone. *Children in Europe*, 14–15.

Vygotsky, L. S. (1933). Uchenie ob emotsijakh. *LS Vygotsky, Sobranie Sochinenij, 6*, 92–318.

Vygotsky, L. S. (1978). *Mind in society: The development of higher psychological processes* (M. Cole, V. John-Steiner, S. Scribner, & E. Souberman, Eds.). Harvard University Press.

Vygotsky, L. S. (1987). *The collected works of L. S. Vygotsky, vol. 1* (R. W. Reiber, Ed.; M. J. Hall, Trans.). Plenum.

Vygotsky, L. S. (1998). *The collected works of L. S. Vygotsky, vol. 5. Child psychology* (R. W. Reiber, Ed.; M. J. Hall, Trans.). Plenum.

Vygotsky, L. S. (2016). Play and its role in the mental development of the child. *International Research in Early Childhood Education, 7*(2), 3–25.

Wells, G. (1999). *Dialogic inquiry*. Cambridge University Press.

Wertsch, J. V. (1985). *Vygotsky and the social formation of mind*. Harvard University Press.

Zebroski, J. T. (1994). *Thinking through theory: Vygotskian perspectives on the teaching of writing*. Boynton/Cook Publishers, Inc.

Zinchenko, V. (2007). Lev Vygotsky: From "silver age" to "red terror". *Children in Europe*, 5–7.

# Chapter 7
# Vygotsky and Secondary-Level Science

## 7.1 Introduction

Chapter 6 focused on preparing young children (4–12-year-olds) to learn about scientific concepts which have been derived from observations and experiments (both lab and thought) since human's earliest musings on the nature of the world around us. One of the earliest known experiments was Eratosthenes' remarkably accurate measurement of the Earth's circumference, nearly 2000 years ago.

Vygotskian theory suggests that young children need to be given opportunities to theorise on natural phenomena at primary school to develop scientific thinking skills (Kravtsov, 2009) for later use. A key example is children generating explanations of phenomena which are *consistent with their observations*. Children's reasoning skills can be developed to a high level towards the end of primary school, given an appropriate set of conditions, including:

- Opportunities to *repeat* an activity so that they can make close observations and to check their reliability,
- *Time to discuss* opportunities to communicate their findings/explanations.
- *Time to consider the link* between their experiment and its broader applications in everyday life and/or in scientific discovery or appliance.

Such preparation in primary school lays the groundwork for children's understanding of scientific concepts to be developed in dedicated science teaching and learning at secondary level.

This chapter presents an action-oriented view of scientific concepts as dynamic, changeable, contextualised and usable *tools* which have been developed over time to help explain the world and how it works. A few years ago, I was invited to produce a chapter for a book entitled *The Future in learning science: What's in it for the Learner*? (Corrigan et al., 2015). My chapter (Murphy, 2015) explored the use and misuse of scientific concepts in secondary-level science teaching. In that chapter (Chapter 7), I suggested that learners and teachers can critique conceptual tools:



C. Murphy, *Vygotsky and Science Education*,
https://doi.org/10.1007/978-3-031-05244-6_7

are they all 'good'? Which ones are fit-for-purpose? Have all been demonstrated definitively to exist? Or are some still models (e.g., the atom)?

The present chapter incorporates ideas from my contribution to Corrigan et al. (2015), together with my research on Vygotsky and the work of his followers on concept development. It also includes research undertaken by one of my doctoral students (section 7.30. The Nugent Study), as well as a Vygotskian perspective on the widely discussed low motivation to learn science at school.

Currently, many scientific concepts are presented in secondary-level school classrooms as 'entities' to be 'learned' and are difficult to transfer to everyday and current science research and practice contexts. Science learning and teaching from a Vygotskian perspective is not organised for the 'average' learner but is structured to support the learning and contributions from diverse students. An ideal learning environment in secondary classrooms sees students and teachers using scientific concepts collaboratively in meaningful contexts, promoting science as asking questions and searching for explanations of phenomena, which is not always straightforward.

The notion of *using*, as opposed to *learning about* scientific concepts, underpins much of the discussion in this chapter. This idea arises from writings of Vygotsky and Wittgenstein. Wittgenstein suggested in his later work that the meaning of concepts lies in their *use* in particular contexts. The idea was also key to the discussions of the 'ordinary language' philosophers of the mid-twentieth century, who suggested that it is more fruitful to *use* high-level concepts (e.g., time) than to try to define them. The lower-level concepts (e.g., a watch, or clock) can then be defined, only because they are able to use the higher-level concept (time) on which it depends.

It is common, however, in science classrooms for teachers to expect students to learn specific definitions of concepts (e.g., energy) outside a context, instead of considering the different ways the *idea of energy* is used both within and outside science. The chapter explores ways that students could be engaged more positively in their learning if teachers loosened the reliance on considering concepts as universal, permanent entities, and highlighted them in their sociocultural contexts.

Table 7.1 contrasts the traditional view of concepts as fixed, universal, and 'true', with the sociocultural, Vygotskian perspective, suggesting that concepts are socially constructed as tools to help explain phenomena: they are subject to change, dependent partly on the technology used to observe, measure and test. In terms of developing concepts, the traditional view suggests that concept development occurs as a consequence of maturation: they can be learned and can replace existing 'misconceptions. The sociocultural view would argue differently that concept development is a dialectical process whereby learners consider both the everyday and abstract concept together during learning, such that the abstract becomes more 'everyday' as the everyday becomes more abstract. Such learning requires an emotional input to generate motivation to ensure that learning can be deep. The third section in Table 7.1 summarises some research carried out on concept development.

**Table 7.1** Moving towards a sociocultural view of scientific concepts
Adapted from Murphy (2015)

| TRADITIONAL | SOCIOCULTURAL |
|---|---|
| **NATURE OF SCIENCE CONCEPTS (SCs)** | |
| SCs exist as entities → | SCs are *created as tools* |
| SCs represent 'truth' → | SCs *created in scientific endeavour* |
| SCs are independent of culture → | SCs are *culture-dependent* |
| SCs are universal → | SCs are *context-bound* |
| SCs are permanent → | SCs are *subject to change* |
| **IDEAS AND THEORIES OF SCIENTIFIC CONCEPT DEVELOPMENT** | |
| Maturation is the driving force of development of SCs – abstract concepts develop later → | *The social world* mediates the development of SCs – even *toddlers use abstract conceptualisation in play* |
| Development leads learning of SCs → | *Learning leads development of SCs* |
| SCs develop linearly → | *SC development is dialectic and occurs via the ZPD. It is NOT linear, but comprises zigzags, gaps, regression, and conflicts* |
| Students' SCs are either correct or 'misconceptions' → | *Students' SC represent their best try at explaining phenomena* |
| SC development occurs when misconceptions are challenged via cognitive conflict → | *SC development occurs via thinking in complexes and pseudoconcepts, towards concepts –* ***not*** *cognitive conflict* |
| SC development is independent of emotion → | *Development of SCs requires emotion* |

(continued)

**Table 7.1** (continued)

| CLASSROOM RESEARCH ON SCIENTIFIC CONCEPT DEVELOPMENT | | |
|---|---|---|
| Teachers create science content for children to learn SCs, based on curricular guidelines | → | *Children learn meaningful science oriented within four major concepts: place, time, materials and conscious reflection* |
| Verbalisation of SCs is assessed as learning | → | *Use of SCs is assessed as learning* |
| Logical SCs 'grow' from experience | → | *Learning of SCs requires bridging into scientific convention* |
| SCs are learned independently, e.g., via IBSE | → | *Learning of SCs is mediated via cultural tools, including language, signs and symbols* |
| Children learn SCs individually | → | *Children learn SCs socially* |
| Learning SCs does not require student dialogue | | *Learning SCs requires forms of dialogue* |
| Learning of SCs is reactive to the teacher | → | *Learning SCs occurs via dialogue and problem-solving with peers and teachers* |
| SCs are 'created' by the teacher for students to learn | → | *SCs are co-constructed by students and teachers* |
| Primary science requires children to learn basic SCs | → | *Primary science provides opportunities to derive scientific explanation from close observation* |
| SCs can be taught over a short period | → | *SCs are developed over a long time* |
| The direction of learning SCs is bottom-up | → | *The direction of learning of SCs is top-down* |

Conceptual learning could be described in Vygotskian terms as the 'leading activity' (Vygotsky, 1967, pp. 15–17) of learning development in secondary-level science. Motivation is key. In this chapter, scientific concepts are explored in terms of their nature and how they can be developed in science lessons, and ways to motivate students in their learning of science. Putting theory into practice is described by presenting a recent study by Nugent (2019) who researched the implementation of social constructivist methodology into her own chemistry teaching practice over a 4-year period.

## 7.2 Social Constructivist Approaches to Science Learning and Teaching

Social constructivism is a theory which suggests that cognitive growth occurs first on a social level, after which it can occur within the individual. Vygotsky's work on CHT (Roth, 2000) suggested that the roots of individuals' knowledge are found in their interactions with their surroundings and other people before their knowledge is internalised. Social constructivism is a useful way forward in science teaching and learning to promote better student engagement and learning in science. Social constructivism became more popular in science education towards the end of the twentieth century. Driver et al. (1994) suggested that scientific knowledge was symbolic in nature and socially developed. They suggested that scientific concepts were:

> *...not the phenomena of nature but constructs that are advanced by the scientific community to interpret nature.* (p. 5)

Rather:

> *...they are constructs that have been invented and imposed on phenomena in attempts to interpret and explain them, often as results of considerable intellectual struggles.* (p. 6)

The sociocultural perspective of scientific concepts summarised in Table 7.1 implies that science lessons should be based on social constructivist approaches, such as active, collaborative learning in which students and the teacher work together in problem-solving situations. Student dialogue and discussion are key aspects of social constructivism, following Vygotsky's argument that language is the most important cultural tool of learning. Implementing social constructivism is a challenge for many school science teachers since it requires a changed mindset, or belief system, and is culturally different from the traditional practice.

Conceptual science can be very difficult for students to learn, but less so when they are taught using social constructivist approaches. The high cognitive demand of SC formation requires time for development of skills and processes. Hence, lack of time, absence from lessons, lack of motivation and disaffection can lead to students struggling to develop SCs, and thus leaving without appropriate understanding, but attaining the pseudoconcepts.

The next section describes a 4-year study which sought to implement and critique the value of social constructivist teaching approaches in the chemistry classroom (Nugent, 2019).

## 7.3 The Nugent Study

Nugent (2019) reported on a study which implemented social constructivist approaches into chemistry teaching over a sustained period of 3–4 years. Her aim was to engage students more closely in their learning, and to improve her own teaching.

The study revealed that a complete teacher mindset change was required, initiated by intensive reading and reflecting on aspects of social constructivism, and implementing methods to mediate complex chemistry concepts and processes, such that students were able to develop and use the concepts more effectively in their learning.

Key changes she made to her teaching were based on developing the ZPD to increase student opportunities to *think at higher levels* by linking their chemistry lessons to real-world science contexts and apprising them of scientific practices to improve meaning-making and by providing more a social and emotional learning environment to develop interest and engagement. More specifically the key changes included:

- Developing and teaching of a new unit, which *encultured students into practices and theories that occurred within the scientific community*, to make scientific practices more meaningful.
- Replacing 'cognitive conflict' with a *dialectical process of communication between teacher and students*, designed to consider information from different views from students and teacher, which were discussed, developed, interpreted, and reasoned.
- *Changing the sequencing* of lessons to provide a more logical, meaningful, and clear organisation, based on student suggestions.
- Introducing new ideas and cultural tools to students, with associated guidance and support, whilst *listening and diagnosing how students were interpreting the activities* and *using this information to inform future practice* – an important learning practice for the teacher.
- Including an *emotional connection* via harnessing student interests, increasing social learning in tasks, using real-life and relevant examples.
- Increasing the prevalence of *active learning*, such that students are leading their own learning as opposed to reacting to the teacher.
- Introducing *cogenerative dialogue*[1] to reduce the power hierarchy, encourage student ownership of their learning, and develop the classroom as a collective.

The activities were tweaked throughout the study. Nugent and the students reflected together upon which and how social constructivist methodologies impacted on teacher and student learning and student engagement. Her data reflected some of the challenges and positives that she felt throughout; for example, the realisation that students were uncomfortable with the 'unknown':

> *I would like to try to de-demonize the unknown for them [the students] a little bit – not to confuse them, but to let them know that the unknown is not a bad thing. It is in fact something that should excite the curiosity within them and understand that science only moves forward because of the unknown.* (Nugent, 2019, Teacher Reflection, p. 94)

[1]A cogenerative dialogue (cogen) is a discourse in which students and teachers participate in a collaborative effort to set in motion positive changes in learning and teaching (Martin, 2009).

After addressing this issue, using examples, peer discussion, pair work and open questioning, she recorded:

> *The test for the 'unknown' worked well and I put an extra challenge on them to try and use the least amount of sample to work out what chemical it was. They had to plan it. Most groups worked very well, and quickly (most completed in a few mins) ... Overheard one of the students say to another 'this is like real science'.* (Nugent, 2019, Teacher Reflection, p. 117)

Student responses to the social constructivist teaching and learning evidenced that they were not afraid to comment on their own and other students' experiences (names are pseudonyms):

> *Steven: I think it's important that like the students know that they need to ask questions cos I think however you explain it, there's going to be something you miss out on... and it's easy to, as a teacher, I'd say it's easy to miss out on something cos like you already know, what you're saying so you can easily miss something that you know but the student doesn't so it's important they can ask the questions.* (Nugent, 2019, Cogen, p. 109)[2]

Through active learning, Nugent was able to diagnose several misunderstandings, such as the structure of organic compounds. For example, students found that building 3D models improved their understanding.

### 7.3.1 Barriers to Implementing Social Constructivist Approaches

The biggest constraint to preparing and implementing widespread change to practice was time, for both teacher and students, particularly in terms of scheme and lesson preparation, and the extended time allocated to student development of understanding. Cogenerative dialogues required planning time, adding another constraint, but were well worth it.

Another potential barrier was managing social learning between students. The teacher needs to help students to develop both their interpersonal skills and emotional intelligence. It was difficult sometimes to balance the support for deeper learning and the need to pass the exam. Finally, student inquiry in terms of experimental design needed to be restricted as the curriculum prescribed specific experiments which were assessed summatively in the exam.

On another level, many barriers exist in terms of policy and implementing policy into practice (Nugent, 2019).

[2] Cogen is an abbreviation for cogenerative dialogue (see footnote 2 in Chap. 4).

### 7.3.2 *Overall Impact of Social Constructivist Teaching Approaches*

Nugent concluded that adopting the social constructivist positively impacted student engagement in the study. The design gradually incorporated students as leaders of their own learning over time. The methodology sought to break down cultural constraints with the focus on learning and formative assessment, rather than exam preparation.

Student dialogue was promoted, and students were more active and engaged than they were in previous classes before the social constructivism was implemented. Peer learning, and the concomitant development of supportive relationships between students, were key. The cogens were vital to opening the teacher-student dialogue, which helped the overall learning environment in the classes, as solutions to learning problems were sought, tried, tested and refined to enhance the learning for all.

Vygotsky's unity of affect and intellect was a guiding principle to enhance the opportunities for mutual comfort and support between students, which they recognised as important in developing their understanding.

The development of a module for students to be more encultured into scientific practice and relate their school chemistry to current and historical issues in scientific research and development developed teacher confidence and created resources which improved student engagement with chemistry. Hearing and acting on student 'voice' improved teacher practice considerably. Developing collaborative spaces for student-student and student-teacher proved to be more effective than asking students for ideas. Importantly, the results of the Leaving Certificate Chemistry exams at the end of the second year evidenced that higher grades were achieved than in previous years' chemistry classes.

Nugent's study (2019) is the result of 4 years' work on the research, theory, and practice of changing teaching practice as a full-time science and chemistry teacher. As such, it exposes a 'warts and all' critique of putting social constructivism into practice. The conclusion indicated that all the hard work was worth it, especially in the unexpected improvement in student performance on summative, national exams. It remains to be seen whether this, and similar studies that evidence increased engagement and achievement of science students via changing from traditional to social constructivist teaching, will become more accepted.

## 7.4 ZPD in Science Learning at Secondary Level

In the previous section, Nugent (2019) spent a lot of time creating ZPDs to support the learning of senior school students in chemistry. Additionally, an important aspect of the ZPD for higher level science students is retention of the learning, which is difficult when there are many demands on the students' attention in several different subjects simultaneously, as well as other interruptions.

Gallimore and Tharp (1990, p. 187) suggested that de-automatisation and recursion occur regularly: "What one formerly could do, one can no longer do." It often happens that self-regulation is not sufficient to restore performance capacity after de-automatisation and thus other assistance is again required (see Chap. 3, Fig. 3.1). Reasons for de-automatisation, suggested by Dunphy and Dunphy (2003, p. 50) in the context of surgical training, could be "environmental change, stress, major upheaval and trauma". In the case of school learning, one major reason for de-automatisation could be due to the variety of different experiences the student has undergone between successive learning situations. De-automatisation as a concept is most helpful in designing ZPDs which allow for cycles of assisted and non-assisted task completion in the development of scientific concepts (see Chap. 9, Sect. 9.3, for more details on regressions and recursion in the ZPD.)

An example of de-automatisation in science learning could be observed in relation to learning basic genetics. Students can be taught all the requisite terminology and the way to perform genetic crosses successfully, but after time and without repeated practice, the ability to perform such crosses needs to be developed anew. Further learning by an individual is made up of these same regulated ZPD sequences, from other- to self-assistance, recurring repeatedly for the development of current and new capacities. There is a mix of other-regulation, self-regulation, and automatised processes for each learner at any one time.

Even the expert can benefit from regulation for enhancement and maintenance of performance (e.g., teachers undergoing CPD).

## 7.5 The Nature of Scientific Concepts

Most definitions of a scientific concept refer to it as an idea, or law, which helps to explain a phenomenon under investigation. For educational purposes, Vygotsky proposed a 'super-concept' framework which defines all human activity within the environment. There are four major concepts in this framework (Kravtsova, 2010):

1. *Time* – all human activity in the world occurs in a certain time.
2. *Space* – all human activity takes place within a space, or place.
3. *Substance* – all human activity uses substance, or materials.
4. *Conscious reflection* – human activity differs from other animals because of the element of reflection on what, how and how to improve, the action or activity.

This framework provides a structure in which every scientific concept can be subsumed. There is a place for the 'process' concepts within Vygotsky's framework under 'conscious reflection'.

### 7.5.1 Theoretical Considerations of the Nature of Scientific Concepts (SCs)

Readers may like to refer to Chap. 3 at this stage, for an introduction to SCs and their use in learning and teaching science. Vygotsky used a model from classical mathematics which suggests that ultimately concepts are all subsumed into one logical system that he calls a 'system of equivalences':

> *The higher levels in the development of word meaning are governed by the law of equivalence of concepts, according to which* ***any concept can be formulated in terms of other concepts in a countless number of ways***. (Vygotsky, 1934/1986, p. 199, emphasis in original)

His broad grid for concepts is based on the surface of a globe, onto which every concept can be placed using a system of coordinates, corresponding to latitude and longitude in geography. A concept's 'longitude' relates to its degree of abstraction, and thus characteristic of thought processes, whilst its 'latitude' represents its objective reference, for example, plant or animal (Vygotsky, 1934/1986, pp. 199–200).

### 7.5.2 Science Concepts (SCs) as 'Tools'

Science education has largely embraced the importance of sociocultural theory as it relates to teaching and learning science; it considers SCs as 'tools' for use in helping to explain and understand scientific phenomena has become more popular. Wells (2008) argued that SCs are not 'possessed' by individuals; rather they provide cultural resources, used for a variety of purposes. SCs can be thought of as 'cultural tools' developed by scientists, to help describe and explain the world around us.

Considering SCs as cultural tools is a much more *active* description. They are constantly being tested for their ability to function as tools in different contexts. Some tools are better than others at doing a specific job. It could be the case that some SCs serve the science context well, but not the science education context – for example, respiration. The term 'respiration' is confused with breathing by younger learners, and the biochemistry of respiration is far too difficult for most senior school biology students to understand, unless they have a good knowledge of chemistry. Some concepts are very tricky to use, especially if there is complex mathematics involved, such as relativity theory.

Are the SCs used in schools for science learning fit for purpose? Or is the question: is the way SCs are taught fit for purpose? It can be useful for students to be made aware that each SC has been generated during the investigation of specific contexts and then replicated to test its generalisability. However, the concept could be more 'robust' if utilised successfully in different contexts.

SCs are not permanent; however, they do change with time and new technologies. Science students might benefit from discussions of how, where and when various SCs came about, including the associated difficulties, political and technological barriers, and enablers, as well as other human factors, to engender a deeper appreciation of the scientific endeavour. The more they understand the process of how SCs are developed will help them to appreciate concepts as tools.

## 7.6 Developing Scientific Concepts (SCs)

The way(s) learners develop SCs has been debated for many years. Conceptual change was very popular in science learning; it was believed that students suffered 'misconceptions' about phenomena and in the process of cognitive conflict when they were challenged with the scientific explanation, they went through a process of conceptual change, drawing on their growing science knowledge and that of teachers and peers (Hewson et al., 1998).

However, conceptual change has not delivered the learning gains needed, for example, many university physics students still misunderstand very basic concepts as evidenced by consistent poor performance on the FCE test (Miller et al., 2013). Miller et al. suggested that unless learners are *using* scientific concepts in, for example, problem-solving, it is unlikely that they will retain the scientific explanations they are presented with after they have learned them for a test or examination. It could be that such scientific concepts are not good for learning out of context and that different pedagogical approaches are needed to be applied to improve the understanding of specific concepts.

### *7.6.1 Concept Development Is Dialectical*

The idea of concept development as dialectical is important at all levels of science learning, so it is relevant in many chapters of this book. Vygotsky (1934/1986, p. 192) proposed a dialectical, as opposed to a linear, model for the development of scientific concepts: "the child's scientific and [her or] his spontaneous [everyday] concepts… *develop in reverse directions* (original italics) … they move to meet each other". For example, the students' everyday concept of *steam* develops more scientifically when they learn about evaporation; at the same time, their concept of *evaporation* will become more everyday to them when applied to familiar contexts, such as *steam, perspiration,* and the phenomenon of *transpiration* in plants (see Chap. 3, Fig. 3.2).

Vygotsky proposed that teachers can *create* a ZDP between the scientific and everyday concepts by illustrating and emphasising the *relationships* between them and showing how the scientific concept can be utilised to explain the everyday concept, whilst simultaneously raising the everyday concept towards its scientific conceptualisation. The task of the teacher, for Vygotsky, is not to evaluate individual conceptions as correct or as 'misconceptions', but rather to help the child, through instruction with respect to the relationships between concepts and within a system of concepts, and to develop conscious awareness and voluntary control of their own thinking (Wells, 2008).

### 7.6.2 The Process of Forming Scientific Concepts in Learning

Vygotsky and his co-workers explored the process of concept formation using a series of *double-stimulation* experiments (see Chap. 4, Sect. 4.2 [ii]). Double stimulation is a principle according to which a subject, when in a problematic situation, turns to external means for support to be able to act (Vygotsky, 1997). The problem is the first stimulus, and the external means is the second stimulus. Vygotsky's double stimulation method placed learners in problem-solving situations that were different from any learning they would have experienced. The experiments thus investigated the formation of *new* concepts via problem-solving tasks requiring the use of non-verbal *signs*. The signs provided a way to solve the problem (Sakharov, 1928).

By the time students reach secondary-level education, many are at the pseudoconcepts stage. This stage is the development of *pseudoconcepts*, which can be confused with true concepts because the learner might be using the right words to describe the concept but lacking the logical connections between its parts. The learner can use the pseudoconcept in communication and activities, such as exams, as if it were a true concept. For example, the learner may use the definition of an ionic bond to describe how it differs from a covalent bond without understanding the nature of chemical bonding.

The words of the learner and teacher may refer to the same idea, but their meanings may not be the same (Gredler & Claytor Shields, 2008). Berger (2005) suggests that true concepts are formed from pseudoconcepts via the *appropriate* use of signs and social (frequently teacher) interventions, thereby forming a bridge between the individual and social meanings. A true concept is bound by logical bonds within parts and between different concepts.

### 7.6.3 The Prevalence of Pseudoconcepts in Science Learning

Many students pass science exams using pseudoconcepts, and only develop the full meaning much later, if at all. The teacher or exam marker may assume wrongly that there is no need for further development.

This confusion between identification of pseudoconcepts and demonstrating understanding of concepts accounts for the common experience of pre-service science teachers that they only begin to understand SCs when they start to teach them. They might have used personally meaningful pseudoconcepts to communicate knowledge successfully using the written form, including the appropriate use of signs, symbols, and scientific terminology. But this may not have been as useful when trying to explain a similar idea without the 'props' of the signs and symbols. It could also explain the experience of tertiary level students, who find that many professors who are experts in their field can give excellent lectures in language they can all understand, whilst academics with less expertise frequently hide behind terminology and complexity.

Eventual formation of true concepts indicates that the learner is now able to master their own thinking. One of the most difficult tasks for learners is to learn the connections and relationships between concepts. It is advisable for teachers and students to construct a large visual diagram of concepts in a topic, and between topics, as the term progresses. Such activity requires pre-planning by the teacher to identify the required concepts for learning in advance.

## 7.7 Motivation to Learn Science

In secondary-level schools the issues facing science teachers are different from those in primary or younger learning contexts. Whereas many primary and early years teachers have problems relating to low teacher confidence to teach science, and lack of support for teachers to promote IBSE. Post-primary science is bedevilled with an outdated, crowded curriculum, assessment of factual knowledge, rigid schemes of learning to be followed by all departmental teachers in some schools, and lack of time and support for IBSE. A consequence of these and other factors in post-primary school has led to many students disengaging from their science lessons. Scientific concept development in many schools follows the traditional approach (see Table 7.1) which is reinforced by the textbook and other resources.

To find out more about student motivation to learn science, I tasked a group of 20 pre-service science teachers with carrying out a quick survey of students in their classes (12–15 years old) to identify which was the most hated topic (SC). The most frequent response was *photosynthesis*; school students said they didn't need it, would never use it, and that it was really boring to learn.

Thinking of photosynthesis as a 'tool for understanding', as opposed to a concept to be learned, caused the pre-service teachers to reflect on how it might be presented to students in such a way that they might be motivated to use it to explain something meaningful to them. Working with the pre-service teachers, we decided upon a more meaningful context for teaching photosynthesis, which might be of interest – perhaps the idea of what can plants do that animal can't?

When the pre-service teachers tried this opening discussion in class, it led to more student interest, especially when it transpired that those plants could make food and oxygen, despite having no brains. Even though most of the students would have learned this already, it was the *context* of that learning – a problem which meant something to them – that motivated them to think: *Well, how can they do that*?

Collaborative investigation of this problem, in which different student groups tackled different sub-questions, led to much more engaged and satisfied responses from students, particularly in tackling more difficult questions, such as how much photosynthesis is needed to sustain the growing human population, which, at the time of writing, is 7.8 billion, and different estimates project it will reach 10 billion between 2083 and 2100. Pre-service teachers reported student-generated questions on deforestation and world food production and distribution arising from these lessons.

This example provides an illustration of the idea that concepts are contextualised, and contexts can be made more meaningful in different ways. Students can be invited to consider why they find certain scientific concepts *difficult*. They can be introduced to ideas as to how scientific concepts are developed, as well as ways in which these concepts were created during the process of scientific investigation.

There are global attempts to improve students' attitudes and motivation towards learning science. *Attitude* is the reaction to science learning, whilst *motivation* is the feeling towards doing such learning. The attitudes of students towards science learning can be influenced by prior perceptions which, in turn, can be influenced by peers, home, and other social and environmental factors.

Archer et al. (2015) developed the idea of 'science capital' as a concept and framework for tracking factors that influence young people's relationship with science. It relates to students' interests, science-related understanding, qualifications, and contacts, particularly those from the home environment. Some researchers argue that students from families with higher science capital (e.g., their parents are scientists or work in scientific fields) tend to have stronger aspirations towards science learning and careers. Others disagree, suggesting gender, science exam performance and confidence as key factors influencing students' attitudes towards science. Indeed, the notion of 'science capital' itself (Archer et al., 2015) has been critiqued by Jensen and Wright (2015) as not being distinct from Bourdieu's bigger concept of *cultural capital*. They maintain that data from Archer et al. (2015) suggests that cultural capital accounts for students' aspirations, and that SC is not a separate construct.

Motivation to study and learn science can be largely determined by self-efficacy – the belief that you will succeed at the task, skill, etc. Success reinforces self-efficacy, which is then believed to enhance performance and achievement. However, continued success can lead to complacency and lower motivation. Other key factors in motivating students to learn science include task value, self-regulation, and the learning environment.

Vygotsky's concept of ZDP can be applied to support and optimise students' learning and achievement. By using the ZPD, teachers, family, and/or fellow learners can help to create conditions which make the learner want to learn for the sake of learning, as opposed to external rewards. Elena Kravtsova, Vygotsky's granddaughter, did a lot of research and development using Vygotsky's CHT. She gave an example of Russian immigrants who might have lived in Germany for 20 years without learning to use the German language. A teenager amongst them watched German cartoons, played clips of interest repeatedly, and hence learned the German required to enjoy the cartoons. This level of German 'incidental' learning enabled him to use the language much more extensively (Kravtsova, 2017). In this context, the main goal of learning German was to create conditions for developing a person's ability to be the subject of their own behaviour, activity and cognition. The key idea for teachers is to develop ZPDs for the learners, such that they can learn science to solve interesting problems, enhance their experience of science lessons, and support their further learning in the area.

One way to try this is to start teaching topics with challenges, beauty or wonders of science; for example, how life began; what was there before the big bang; how and when were the elements formed; why the Moon is escaping the Earth; and how and why plants grow in so many different shapes. A team of physicists, science educators and web developers created what turned out to be an engaging website aimed at public and school student engagement. The website initially attracted people via 'adverts' on the local Dublin train service (called the Dublin Rapid Area Transit service, DART) and sustained the interest with an intense social media campaign that generated a city-wide conversation about physics (see Burke, 2013). The website was used mostly for supporting physics learning and teaching in post-primary schools. The work of Lancaster et al. (2015) also stressed the link between 'real' science and school science as essential to future science learning. The role of the current teacher in making school science more engaging is discussed by Loughran and Smith (2015).

## 7.8 Summary and Conclusion

In terms of learning and teaching scientific concepts, there are many differences between the traditional approach, which is individually centred, and the sociocultural approach, which has been addressed earlier in the chapter. These differences are summarised in Table 7.1, to provide guiding thoughts for teachers, learners, researchers, curriculum developers and other stakeholders in science education as ways to reconceptualise scientific concepts in science learning and teaching to make it more pleasurable, challenging, and, in the long-run, more effective.

In conclusion, this chapter provides a theoretical and practical exploration of scientific concept development and motivation to learn science. It features a longitudinal, 4-year study which implements and critiques the use of social constructivist teaching and learning approaches in school chemistry classes.

The aim of the chapter is to provoke discussion and interest in looking at traditional science learning a bit differently. If teachers look at scientific concepts more critically as tools for science teaching (e.g., as in the photosynthesis and dartofphysics examples described above) they can make science lessons more engaging for themselves and for students.

The move towards more collaborative and cooperative learning strategies in which students are encouraged and facilitated to repeat experiments as required mimics more closely the science world that they may wish to enter. The Nugent (2019) study describes the reality of attempting this in practice. Essentially, the move from teaching the curriculum towards teaching students by engaging their interests and relating that work to the curriculum, may improve significantly their scientific concept development and motivation to learn science subjects.

> *By giving our students practice in talking with others, we give them frames for thinking on their own.* (Vygotsky, 1978)

## References


Archer, L., Dawson, E., DeWitt, J., Seakins, A., & Wong, B. (2015). "Science capital": A conceptual, methodological, and empirical argument for extending bourdieusian notions of capital beyond the arts. *Journal of Research in Science Teaching, 52*(7), 922–948. Retrieved from https://doi.org/10.1002/tea.21227

Berger, M. (2005). Vygotsky's theory of concept formation and mathematics education. In H. L. Chick & J. L. Vincent (Eds.), *Proceedings of the 29th conference of the international group for the psychology of mathematics education* (pp. 153–160). University of Melbourne.

Burke, E. (2013). *Commuters get a DART of physics education on train routes*. Retrieved from https://www.siliconrepublic.com/author/eburke

Driver, R., Asoko, H., Leach, J., Mortimer, E., & Scott, P. (1994). Constructing scientific knowledge in the classroom. *Educational Researcher, 23*, 5–12.

Dunphy, B. C., & Dunphy, S. L. (2003). Assisted Performance and the zone of proximal development (ZPD): A potential framework for providing surgical education. *Australian Journal of Educational and Developmental Psychology, 3*, 48–58.

Gallimore, R., & Tharp, R. (1990). Teaching mind in society: Teaching, schooling, and literate discourse. Vygotsky and education. Instructional implications and applications of sociohistorical psychology. In L. Moll (Ed.), *Vygotsky and education* (pp. 175–205). Cambridge University Press.

Gredler, M., & Clayton Shields, C. (2008). *Vygotsky's legacy: A foundation for research and practice*. Guildford.

Hewson, P. W., Beeth, M. E., & Thorley, N. R. (1998). Teaching for conceptual change. In K. G. Tobin & B. J. Fraser (Eds.), *International handbook of science education* (pp. 199–218). Kluwer.

Kravtsov, G. (2009). *A cultural-historical basis of the "golden key" program*. Paper presented at the International Psychological Vygotsky Summer School, Belaya Kalitva, Russia.

Kravtsova, E. E. (2010). The cultural-historical foundations of the zone of proximal development. *Journal of Russian and East European Psychology, 47*(6), 9–24.

Kravtsova, E. E. (2017). The sense and the meaning of cultural-historical theory of L. S. Vygotsky. *Revue Internationale du CRIRES: Innover dans la Tradition de Vygotsky, 4*(1), 35–47.

Lancaster, G. Panizzon, D., & Corrigan, D. (2015). Pursuing different forms of science learning through innovative curriculum implementation. In D. Corrigan, C. Buntting, J. Dillon, A. Jones, & R. Gunstone (Eds.) *The future in learning science: What's in it for the learner?* (pp. 101–126). Springer. Retrieved from https://doi.org/10.1007/978-3-319-16543-1

Loughran, J., & Smith, K. (2015). Facilitating change in science teachers' perceptions about learning and teaching. In D. Corrigan, C. Buntting, J. Dillon, A. Jones, & R. Gunstone (Eds.) *The future in learning science: What's in it for the learner?* (pp. 279–294). Springer. Retrieved from https://doi.org/10.1007/978-3-319-16543-1

Miller, K. A., Lasry, N., Chu, K., & Mazur, E. (2013). Role of physics lecture demonstrations in conceptual learning. *Physical Review Special Topics – Physics Education Research, 9*(2), 1–5.

Murphy, C. (2015). *Reconceptualising the learning and teaching of scientific concepts*. In D. Corrigan, C. Buntting, J. Dillon, A. Jones, & R. Gunstone (Eds.), *The future in learning science: What's in it for the learner?* (pp. 127–150). Springer. Retrieved from https://doi.org/10.1007/978-3-319-16543-1

Nugent, S. (2019). *Exploring the use of social constructivist methodologies in teacher practice and student engagement in chemistry*. Unpublished DEd thesis, Trinity College Dublin.
Roth, W. M. (2000). Artificial neural networks for modelling knowing and learning in science. *Journal of Research in Science Teaching: The Official Journal of the National Association for Research in Science Teaching, 37*(1), 63–80.
Sakharov, L. (1928). *Methods for investigating concepts*. Retrieved from http://www.marxists.org/archive/vygotsky/works/comment/sakharov.htm
Vygotsky, L. S. (1934/1986). *Thought and language*. MIT.
Vygotsky, L. S. (1967). Play and its role in the mental development of the child. *Soviet Psychology, 5*(3), 6–18.
Vygotsky, L. S. (1978) *Mind in society: The development of higher psychological processes* (M. Cole, V. John-Steiner, S. Scribner, & E. Souberman, Eds.). Harvard University Press.
Vygotsky, L. S. (1997). *The collected works of L. S. Vygotsky, vol. 4. The history of the development of higher mental functions* (R. W. Reiber, Ed.; M. J. Hall, Trans.). Plenum.
Wells, G. (2008). Learning to use science concepts. *Cultural Studies in Science Education, 3*(2), 329–350.

# Chapter 8
# Vygotsky and Informal Science Learning

## 8.1 Introduction

The impact of science on society will only reach its full potential when science becomes integral to daily conversations. The focus of this chapter is on bridging the informal and formal setting of science learning to explore the potential to provide opportunities for *talking science*, via *spontaneous*, rather than *reactive* learning. Most formal science learning involves students reacting to instructors' questions or instructions and is individually focused. Informal science learning, on the other hand, offers a choice of experiences from which spontaneous learning comes from voluntary engagement via interest, and is largely social. The research highlighting the importance of talking science, whether through conversation, dialogue, argument, etc., is compelling (for example, Driver et al., 1994). The previous three chapters on science education in formal contexts foregrounds the importance of social constructivism in science learning. The informal context holds much more potential to embrace the social aspects of learning conceptual science.

Informal, also referred to as non-formal science education, could be described as learning in science *beyond* school, such as science learning experiences in museums and science centres, or indeed students' experiences of science clubs, and science competitions *within* school. There is an increasing awareness of the value of informal science learning, both in the 'informal sector' (e.g., Stocklmayer et al., 2010) and in school (e.g., Dillon, 2013). Vygotsky highlighted how formal school learning divorced from the real world could be highly ineffective. He argued:

> *...that the school has been locked away and walled in as if by a tall fence from life itself has been its greatest failing. Education is just as meaningless outside the real world as is a fire without oxygen, or as is breathing in a vacuum.* (Vygotsky, 1997, p. 345)

Science itself needs to become more *inter-disciplinary* so that implications of scientific research and teaching can be implemented more effectively in societal contexts. Scientists can learn from the worlds of sport, current-affairs, and the arts, where platforms and forums bring together rich, diverse groups, using media to facilitate



C. Murphy, *Vygotsky and Science Education*,
https://doi.org/10.1007/978-3-031-05244-6_8

conversations and elicit spontaneous learning in these topics. The term 'transdisciplinary science' relates to specific issues, such as climate change and pandemics. Such issues require research, knowledge, and methodologies from scientific, social, and cultural perspectives to name a few, which, with the collaboration of a wide diversity of scholars, all focus on addressing the issue at hand.

Vygotsky emphasised the importance of conversation in that greater learning is achieved in collaboration than by individuals working alone (Vygotsky, 1986). He proposed that the collaboration of individuals who have experienced transformative learning experiences can lead to wider societal transformation. More recently, the work of scientists, Meltzoff et al. (2009) carried out an extensive, interdisciplinary project, bringing together some of the best research in education, psychology, neuroscience, and artificial intelligence. Their findings revealed that the key component determining how people learn is the importance of the social, and how social interaction via conversation is a "powerful catalyst for learning" (p. 288).

Vygotsky's identification of reactive and spontaneous learning (Kravtsova, 2009, personal communication) suggested that learners follow their own programme in spontaneous learning, driven by interest and curiosity. Students spend nearly 80% of their waking hours outside school: they learn at home, online, in community-centres, clubs, at science-centres, museums, and through digital media and gaming. Informal science contexts can spark curiosity and conversation, as well as eliciting spontaneous scientific learning outside school.

Combining the key elements of *talking science* and *spontaneous learning* of science, highlights potential synergies between informal and formal science learning to produce scientifically more fluent citizens.

Some of the key extreme characteristics of informal and formal learning are summarised in Table 8.1.

A Vygotskian perspective on the roles of formal and informal science learning embrace two of his major constructs: the dialectical relationships between *everyday* and *scientific concepts* and between *informal* and *formal* science learning and linking these dialectical relationships in the *ZPD*. The chapter continues with a consideration of Vygotskian ideas, focusing on the dialectical relationship between informal and formal science learning, the ZDP, and the role of imagination in science learning. The key finding is that informal and formal science learning should be acknowledged as inseparable for effective science learning.

## 8.2 Dialectical Relationship Between Informal and Formal Science Learning

Science learning does not occur only in school or other formal learning contexts, although it is perceived that the formal context appears to be the most important. Formally learned science is assessed mostly by examinations, and these results can determine whether a student is suitable for a science-related career.

**Table 8.1** Characteristics of formal and informal learning

| Formal | Informal |
|---|---|
| School curriculum focus | Less structured activities |
| Extrinsic motivation | Exploration, experimentation, intrinsic motivation |
| Strict assessments, measuring specific learning outcomes | Learning outcomes not explicitly foregrounded and less formal assessment |
| Decontextualised, more explicit knowledge | Contextualised and more tacit knowledge |
| Non-authentic | Authentic |
| Less verbalisation | More verbalisation |
| Little interaction with scientists | Potential for strong interaction with scientists |
| Little space for creative thinking and wonder | More opportunities for wonder, curiosity and creative thinking |
| Opportunities for problem-solving | Perhaps less opportunity for problem-solving? |
| Potential for formation of pseudoconcepts | Potential for formation of pseudoconcepts |

There is currently a growing demand for more students to enter broader science, considered as STEM or STEAM careers. STEM represents the subjects of Science, Technology, Engineering and Mathematics. STEAM is STEM with the important addition of Arts. However, but the demand is not being met in many countries. There are several barriers that contribute to the issue of lack of uptake of science subjects at higher levels in schools and in third-level institutions. A report from the Institution of Engineering and Technology (2008) identified four major barriers:

- The need for quality teaching for students to become, and remain, engaged in STEM/STEAM.
- The perceived difficulty of STEM/STEAM subjects.
- The disillusionment of the transition from primary to secondary school.
- The negative views about success in, and 'unacceptable' stereotypes about STEM/STEAM.

The same report identified mitigating factors, in which informal science learning contexts, such as out-of-school, hands-on, challenging programmes, offer high-value prizes for the *students* as opposed to the schools. They suggested that such programmes, offered in diverse STEM/STEAM areas, could alter student beliefs about opting for easier subjects, and enhance self-efficacy.

There have been calls to bridge formal and informal science learning (e.g., Hung et al., 2012; Leonard et al., 2017). Vygotsky implied that there is no gap between them anyway, in his dialectical description illustrated above, in which the ZPD is expanded. Brown et al. (1993, p. 191) expanded the ZPD to include "people, adults and children, with varying levels of expertise, but it can also include artifacts, such as books, videos, wall displays, scientific equipment and a computer environment needed to support intentional learning".

The ZPD concept could be extended further to offer more learning opportunities in terms of promoting dialogue with scientists and experiencing science as it happens, and a focus on the affective aspects of science and scientists, in environments such as Science Gallery. Science Gallery is an international group of public science centres, developed from a concept by a group connected to Trinity College Dublin, Ireland. Communication within and between Science Galleries can be facilitated by Twitter, the platform which enables direct contact between learners, teachers, and scientists.

Vygotsky's theory foregrounded the interaction of the use of 'tools' and 'symbols' in the roles played by participants in the learning process. Hence the idea of "divergent classrooms" as learning communities in which each participant makes a significant contribution to an emergent understanding between them, despite having unequal knowledge (Palincsar et al., 1993, p. 43). Scott and Mortimer (2005) researched student dialogue in science classrooms, and suggested that such dialogue can result in meaning-making.

In a large study of different forms of informal science learning, informal science experiences at home (e.g., science kits, TV and other media), visits to science centres, etc., away from home, outdoor nature experiences (e.g., forest parks, lakes), and semi-formal science experiences (e.g., summer camps, after-school science clubs), Lin and Schunn (2016) concluded that there are unique benefits from students' informal science learning experiences across the different forms. They suggested that each form offers affordances, and that the challenge is to create equitable opportunities for students to experience the broadest possible access to informal and formal science learning.

## 8.3 The ZPD in Informal/Formal Science Learning

For the purposes of this chapter, the ZPD can be considered as activity that is dynamic and fluid. The goal is one of unity of both informal and formal science learning, keeping in mind that the individual and society should both benefit; it can be suggested that formal and informal learning of science are interdependent. The focus is on the creation of multiple ZPDs within an asymmetrical framework, including progression and regression. The idea is to create new spaces to bridge informal and formal science, which, ideally, could become a new way of learning science, and thus a new community of science learners and learning.

The ZPD is often defined as "the distance between the actual developmental level as determined by independent problem solving and the level of potential development as determined through problem solving under adult guidance or in collaboration with more capable peers" (Vygotsky, 1978). The following points may be useful for application to the creation of ZPDs to integrate formal and informal science learning:

- The ZPD deals with each learner's *potential*. Thus, activities which are aimed at bridging informal and informal need to be planned in a way that both settings contribute towards the agreed potential, which is being developed by learners within specific activities, for example, problem-solving.
- The ZPD relates to functions that are not yet emergent (Robbins, 2003). Robbins gives a useful example of the way adults react to a baby in their use of baby talk. They use both baby talk and 'adult' talk in the expectation that the infant will grow into the community and one day be able to use similar language. This metaphor demonstrates how teachers need to challenge learners with their *potential* in mind, as opposed to their current level of development.
- The ZPD should also focus on ways to promote the learner's personal transformation, which might be in the form of acknowledging regression, as they realise that learning progresses more like the tide (backwards and forwards, but continuing in the forward direction), as opposed to progression in a straight line upwards.
- The ZPD provides a focus on the process, not the product of learning, since learning is a continuous process in that it leads to more learning, especially if teachers promote learning via both formal and informal pedagogies.
- The *internalisation* aspect of the ZPD relates to changes in the learner themselves, in terms of the way they think and reflect on themselves and the world around them. The combination of learning in different settings provides more potential for the internalisation of knowledge and processes.
- The ZPD is not only a cognitive zone – it includes play, creative imitation, and hence the potential for reshaping thinking and learning in new ways. It is thus considered as emancipatory, rather than fixed.
- The ZPD provides for new connections, categories and end points – providers can connect new knowledge to old, generating new categories, to help imbue a broader understanding by students and other learners.

A useful definition of the ZPD was offered by Lois Holzman (1997, p. 171):

> *A ZPD is a form of life in which people collectively and relationally create developmental learning that goes beyond what any individual in the group could learn on her or his own. Our effort is to create continuously overlapping ZPDs, a particular relational activity that simultaneously is and makes possible the transforming of rigid behavior (forms of life that have become alienated and fossilized) into new forms of life.*

### *8.3.1 Example of the ZPD in Informal/Formal Science Learning*

Researchers and teachers from various disciplines and from public and private sectors developed a project to bridge informal and formal science education, with a focus on physics. We developed a series of short scientific adverts on the Dublin metro system, the DART. Each statement was linked to a custom-built website. Social media and traditional media directed commuters to the 'DART of Physics' campaign and sustained their interest once initially engaged (see, for example, Fig. 8.1). The project developed into a national campaign to engage the public, as well as students, in the fascination of physics.

The idea was to generate conversations about physics within the public via commuters, the media, and educational institutions. The key project co-ordinators were science educational theorists and informal science practitioners. The main proposition was that informal science learning promotes conversations, discussion, and question-posing about scientific phenomena.

Formal science learning helps students acquire knowledge about scientific concepts, their applications and some problem-solving, but does not promote sufficient conversation or dialogue in the classroom. Bridging the two forms of science learning via the ZPD can lead to better meaning-making and hence an improved experience of learning, which in turn leads to better science learning outcomes. In this project, the ZPD comprised the website, adverts, a publicity campaign and use of social media.

**(a) Physics advert**

**(b) commuter interaction via QR code**

**Fig. 8.1** (**a**) Physics advert (**b**) commuter interaction via QR code

### 8.3.2 Outcomes for Science Learning Outside School or University

DART of Physics directly connected researchers/physicists with the public. It allowed their passion and personalities to come through. Most science communication tries to connect science to the public through the 'everyday' approach – e.g., how does your microwave work, what products can nanotechnology offer you.

DART of Physics took a different approach: it appealed to the curiosities, to the imaginations and to the creativities of the public – the values that make individuals amazing scientists. DART of Physics aimed to provoke a conversation between physicists and the public – one on an equal playing field where dialogue trumped didactic approaches. An image of one of the adverts from the DART of Physics campaign to the Facebook page of 'I F***ing Love Science', which was approved by them (not an easy process). It was viewed by approximately *9.4 million users.*

Additionally, more than 22,000 Facebook users showed their approval for the image by 'liking' the image, with a further 1649 leaving comments. A link was provided to the DART of Physics website with a surge of over 10,000 new visitors recorded in less than 12 h. These website figures would have been higher, but the colossal interest crashed the site for a brief time.

Members of the public were interviewed on the DART, or as they alighted the trains. While the majority were positive, some had failed to notice the adverts. There was no test to measure science knowledge amongst the commuters: just what they thought of the campaign. There were no negative comments. A couple of examples of Dubliners' comments were:

> *I think it was great a creative campaign, good for [keeping my] attention, good for marketing, and really got people thinking about physics* (from a man, mid 50s).

> *Adverts like this are important because you become more knowledgeable, and you can maybe integrate different things in a more interactive way* (from a woman, 18–30).

### 8.3.3 Outcomes for Formal Science Learning in School

Research on the potential impact of DART of Physics on formal science learning was carried out by pre-service science teachers, who investigated the effect of using the DART of Physics adverts and the website on student engagement with physics. They prepared teaching and learning resources for classes and reflected afterwards that their own learning had improved as much as that of their students. An indicative quote was:

*DART of Physics statements were all heavily focused on building a connection between the scientific world and the everyday world and as a result students responded positively to the statements.*

In an all-girls school, one of the questions asked by the pre-service teacher was whether the DART of Physics lessons made them think differently about physics, and typical responses were:

- *Yes, I didn't realise physics is used in the outside world.*
- *Yes, because when you think of physics it seems boring, but it's interesting.*

The first of the above two comments from students highlights a problem with school science, in that much of it stays in the classroom, with no link to the outside. Some of the lessons learned by the pre-service teachers from their DART of Physics teaching experiences were:

- *Previous research has demonstrated that students desire more opportunities for practical work, extended investigations, and opportunities for discussion in science… I incorporated all three of these activities into my lesson plan and noted a marked improvement in student enjoyment, participation, and positivity during the class.*
- *The science curriculum [needs to] be updated to focus on modern advances in science and student led investigations.*
- *Students' negativity with school stems from further than the school gates and thus the challenges of introducing interactive teaching are there for both students and teachers.*
- *The importance of engaging, hands-on activities, whilst giving students responsibility and resources to explore and investigate.*
- *Science education needs to be relevant to everyday life, contain student-driven investigations and display positive attitudes from the teacher… field trips encouraged … career prospects must be highlighted. This will further promote science in the real world and may inspire young people to study and pursue a career in science.*

This example of a ZPD shows that informal and formal science, when combined to harness a more holistic approach to learning science, holds the potential for much more effective science pedagogy. A key aspect of the DART of Physics example was bringing the local element into students' physics lessons, particularly in the comment above from the child who was surprised that there was physics in the "outside world". So many linkages between school science and science itself need to be enhanced, and ZPD creation, which consciously aims to do this via bridging informal and formal science learning experiences, could be an effective way forward.

### 8.3.4 Example of Linking Industry with Primary Schools

Another approach to linking informal science experiences with school learning can be seen in the STEAM-in-a-Box (SIAB) project, (discussed in more detail in Chap. 6). SIAB provides a framework for industry and scientific professionals to coteach science, technology, engineering, arts and mathematics with primary-school teachers (see Sect. 6.3 (d) for more information). The science professionals and primary teachers undergo induction in coteaching pedagogy (Murphy, 2016), which aims at supporting them to share expertise in providing a stimulating, exciting, real-science learning experience for children.

The vision of STEAM Education Ltd is to inspire young children to become the next generation of scientists, technologists, engineers, artists and mathematicians. It develops innovative, fun and engaging educational resources in these subject areas specifically for upper primary schools. It facilitates coteaching partnerships of science and arts industry professionals and academic experts with teachers to deliver these programmes, multiplying the benefits to all actors involved: the children, the teachers and the outreach experts. SIAB was the brainchild of a parent-scientist, who wanted to try out teaching ‘rocket science’ to children in 5th or 6th class (children aged 10–12).

The programmes run for an hour per week over a period of between 10 and 25 weeks. STEAM-Education-Ltd (https://www.steam-ed.ie/about-us/) comprises a unique partnership that unites actors from STEAM research, science education research, formal and informal science education, artists, designers and industry with one vision – to excite, inspire, and educate primary school children in STEAM through a direct connection with frontier research and development, via a partnership between industry, schools and academia. It also provides additional resource materials, specially designed for class teachers, and is developing a CPD programme in STEAM education. The framework seeks to make a step change in STEAM education in Ireland through new investment and by leveraging existing resources. There is *no assessment of children* – their experience in these classes is to promote their enjoyment and interest in science learning.

The STEAM coteacher arrives at school once a week with a box of scientific materials, supported with engaging PowerPoint presentations and videos, and for an hour the coteachers engage the children in ‘rocket science’. The programme content was designed to enhance the current primary school curriculum, while complementing and leading into the new Junior Cycle science programme. It addresses the nature of science as an overarching strand, and moves through the physical, chemical, Earth and space and biological strands, using the ‘big history’ of the universe as both a means of structuring the course and as a narrative device. Children receive a STEAM journal at the beginning of the programme and are encouraged to follow up questions that arise in the classroom at home, do their own research and attempt to find the answers themselves. Taking STEAM home with them is one way of extending their *ZDP* by encouraging reflection on their learning, collaboration with friends and family and linking STEAM between home and school. The focus is on children’s experience, and assessment is via quizzes and games.

The sharing of ideas, experience and expertise lies at the root of coteaching, as STEAM professionals and primary school teachers work together in co-planning, co-practising and co-evaluating for a series of approximately 10 lessons. Coteaching develops both coteachers' confidence as they share expertise and co-reflect on their progress towards providing 'ideal' learning environments for children. George Bernard Shaw's words illustrate the difference between sharing resources and sharing ideas, or expertise.

> *If you have an apple and I have an apple and we exchange these apples, then you and I will still each have one apple. But if you have an idea and I have an idea and we exchange these ideas, then each of us will have two ideas.* (George Bernard Shaw, see below for more information)[1]

Further development of this framework is planned to increase engagement and input from, and between, different levels of the educational ecosystem and industry as leading to an "integrated educational ecosySTEM" (see Fig. 8.2).

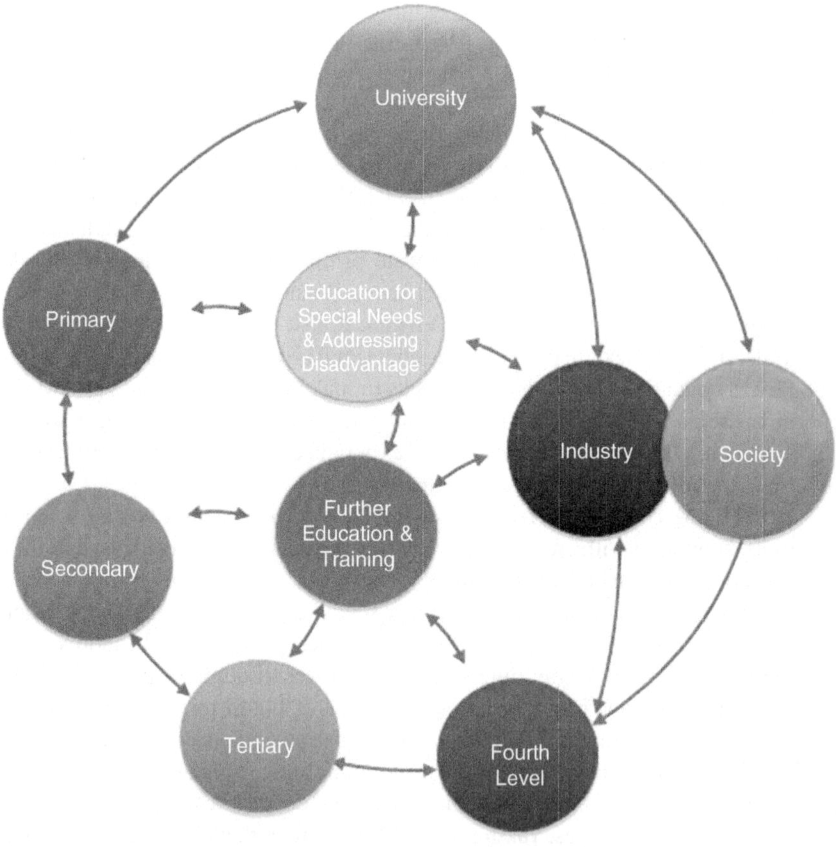


**Fig. 8.2** The STEAM integrated educational ecosySTEM framework

[1] https://www.goodreads.com/quotes/23088-if-you-have-an-apple-and-i-have-an-apple

Research on coteaching science in primary schools shows that extraordinary results can be obtained through external specialists working closely with the cooperating classroom teacher (Murphy, 2016). Coteaching via shared expertise provides a pedagogy which can be used to promote both teacher and student development of twenty-first-century learning skills, which include:

- Critical thinking and problem-solving.
- Collaboration across networks.
- Curiosity and imagination, empathy.
- Persistence.
- Grit.
- Global stewardship.

In addressing these needs, the SIAB programmes provide a sustainable solution to improving science learning, and setting the learning in global contexts, for example, climate change education. The goal is to harness and share expertise via this public-private-industry collaboration to improve the STEAM learning of all students at every primary school; and thereby to increase the diversity in STEAM fields and the STEAM literacy of the Irish nation and beyond. The SIAB programmes also set out an ambitious programme of research through practice that will have high impact and will be transformative in the science curriculum in Ireland, with further opportunities for a global impact.

Evaluation of the impact of the programmes is the next focus, to capture the elements of SIAB which are most effective in inspiring children in STEAM areas and introducing them to the place of STEAM in society. So far, evaluations from two master's theses researching the early years of SIAB have found that the programme thus far is successful in terms of child and teacher positive attitudes towards the lessons and highlight the importance of the coteaching element. However, not all coteaching pairs work as successfully as others, although even the less effective coteaching teams suggest that the coteaching is key to enhancing the success of the programme. Another key focus is to expand into new school programmes, including health and well-being and climate change.

Earlier in this book, informal and formal science learning are positioned in a dialectical model, designed to bridge the gap between them (see Sect. 3.4, Fig. 3.3). As formal science teaching and learning become more informal, and vice-versa, there is more potential for synergistic outcomes, in which informal science environments provide more scope to create and build on formal science learning. At the same time, formal science contexts can introduce more informal learning situations, which facilitate dialogue, such as role-play and the use of imaginary situations in science lessons.

## 8.4 The Role of Imagination in Science Learning

Another area in which closer links between informal and formal science learning are required is in the importance of imagination in all areas of science, particularly in concept development.

In science, abstract concepts are imagined initially, and then used as 'tools' in the real world to explain natural phenomena. In formal science learning contexts, abstract concepts are presented as entities to be learned (memorised) and are difficult to transfer to everyday and current science research and practice contexts. It is common in science classrooms for teachers to expect students to learn specific definitions of concepts (e.g., energy) outside a context, instead of considering the different ways the *idea of energy* is used both within and outside science.

Science learning from a Vygotskian perspective posits that learners and teachers use scientific concepts *collaboratively* in meaningful contexts, promoting science as asking questions and searching for explanations of phenomena, which is not always straightforward. Firestein (2013) describes the process of scientific endeavour as akin to looking for a black cat in a dark room, and there may not even be a cat in the room. His advice for getting the feel of how scientists think is:

> *Next time you meet a scientist – at a dinner party, at your child's school, just by chance – don't ask her to explain what she does. Ask her what she's trying to figure out.* (p. 82)

Conceptual learning in science from a Vygotskian perspective is not effective via the use of transmissive learning, which leads to the development of pseudoconcepts, as opposed to true concepts. Pseudoconcepts can be confused with true concepts because the learner might be using the right words to describe the concept but lacking the logical connections between its parts. A true concept is bound by *logical* bonds within parts and between different concepts (Murphy, 2015). Developing true concepts requires contextualisation, both in terms of linking concepts to how they are used in different areas of science research and development.

Equally important is to provide the student with the 'story' of how and when specific abstract concepts were developed, such as the double-helix structure of DNA suggested by Watson and Crick (1953) and the idea of semiconservative replication of DNA as a requirement for inheritance of genetic material, or the Blackman (1905) experiments which led to his law of limiting factors determining the rate of photosynthesis.

### *8.4.1 Imagination and Play in Science Learning*

Imagination is key to science – how else do scientists and teachers come up with explanations for phenomena such as black holes, galaxies, and methane bubbles? Human imagination develops from birth, particularly in play contexts. Playful behaviour has highly positive effects on cognitive brain functioning. Vygotsky proposed the two defining characteristics of play as the *imaginary situation* and *rules*. He said that all play creates an imaginary situation, and all imaginary situations contain rules, which stem from the imaginary situation (see Ackerman, 1999). In *game play*, the rules are overt, whilst in *free play*, the imagination dominates.

Neuroscience experiments have shown positive effects of play on the brain, for example, early experiments comparing the brains of rats raised in an 'enriched' environments, in exciting, toy-filled colonies and 'impoverished' rats in boring, solitary confinement (Diamond et al., 1964). They showed that the rats raised in resource-rich environments had thicker cerebral cortices. Later, Greenough and Black (1992) performed similar experiments and demonstrated that the rats raised in resource-rich environments were also more intelligent, as measured by their success rates in navigating mazes. After bouts of rough and tumble play, Gordon et al. (2003) showed that rat brains evidenced increased brain-derived neurotrophic factor (BDNF) levels – BDNF is essential for growth and maintenance of brain cells. Other work showed that BDNF levels were higher in rats that had been allowed to explore (Huber et al., 2007).

Educational research indicates that short, unstructured breaks increase attention to academic activities (Pellegrini & Holmes, 2006), but the same effect is not evident if the break is used for structured physical exercises. Role-play, as described above, improves conceptual learning. Cowles (2017) drew the connections between elementary learning by young children and scientific theorising. He maintained that science has always been child's play and drew attention to Dewey's (1910) *How We Think* book in which his short schematic of children's learning became the axiomatic modern representation of scientific thought. Dewey suggested that:

> *Upon examination, each instance reveals, clearly, five logically distinct steps:*
>
> (i) *a felt difficulty;*
> (ii) *its location and definition.*
> (iii) *suggestion of possible solution.*
> (iv) *development by reasoning of the bearings of the suggestion.*
> (v) *further observation and experiment leading to its acceptance or rejection; that is, the conclusion of belief or disbelief.* (Dewey, 1910, p. 72)

This argues Cowles (2017) is the modern 'scientific method', despite Dewey's intention to describe 'ordinary' thinking. The closeness of children's thinking and that of scientists could be boiled down to two features, spontaneity, and sociality. It could be argued that creating a way forward for science education in both informal and formal contexts (ZPDs), which promote both spontaneity and sociality in learning, points the way forward for improving science learning – anywhere.

## 8.5 Summary and Conclusion

This chapter has considered ways to bridge informal and formal science learning to enhance the process of learning science at all levels. The formal context lacks sufficient opportunities for learners to discuss science topics in detail; to 'play' with ideas in collaboration with peers; or to use their imagination. The informal sector, on the other hand, lacks sufficient opportunities for learners to link their experiences to the science they have studied, or will be studying in the more formal contexts. The Vygotskian dialectical model of informal and formal science learning predicts that making formal science learning more informal, and vice-versa, opens opportunities to lessen the existing gaps between them.

Examples of projects and other innovative ideas which attempt to bridge the informal-formal science gap have been given in the chapter. Perhaps the most sustainable example is the SIAB programme, which links science industries with primary schools to provide a more exciting learning environment for the children, and in which the schoolteacher and scientist coteacher plan, teach and evaluate a series of lessons (usually 10-week or 25-week programmes which involve an hour of coteaching per week). This model directly addresses the Vygotskian dialectical model and, at the same time, promotes a social constructivist teaching approach and creates an expanded ZPD for the learning of all participants.

## References

Ackerman, D. (1999). *Deep play*. Random House.

Blackman, F. F. (1905). Optima and limiting factors. *Annals of Botany, 19*(2), 281–296.

Brown, A. L., Ash, D., Rutherford, M., Nakagawa, K., Gordon, A., & Campione, J. C. (1993). Distributed expertise in the classroom. In *Distributed cognitions: Psychological and educational considerations* (pp. 190–191, 201–204). Cambridge University Press.

Cowles, H. (2017). Child's play. *Aeon Magazine*. Retrieved from https://aeon.co/essays/how-the-scientific-method-came-from-watching-children-play

Dewey, J. (1910). *How we think*. DC Heath.

Diamond, M. C., Krech, D., & Rosenzweig, M. R. (1964). The effects of an enriched environment on the histology of the rat cerebral cortex. *Journal of Comparative Neurology, 123*(1), 111–119.

Dillon, J. (2013). Doing justice to the informal science education community. *Studies in Science Education, 50*(1), 131–136.

Driver, R., Asoko, H., Leach, J., Mortimer, E., & Scott, P. (1994). Constructing scientific knowledge in the classroom. *Educational Researcher, 23*, 5–12.

Firestein, S. (2013). *Ignorance: How it drives science*. Oxford University Press.

Gordon, N. S., Burke, S., Akil, H., Watson, S. J., & Panskepp, J. (2003). Socially induced brain "fertilization": Play promotes brain derived neurotrophic factor transcription in the amygdala and dorsolateral frontal cortex in juvenile rats. *Neuroscience Letters, 341*(1), 17–20.

Greenough, W. T., & Black, J. E. (1992). Induction of brain structure by experience: Substrates for cognitive development. In M. R. Gunnar & C. A. Nelson (Eds.), *Minnesota symposia on child psychology: Developmental neuroscience* (Vol. 24, pp. 155–200). Lawrence: A Erlbaum Associates.

Holzman, L. (1997). *Schools for growth: Radical alternatives to current educational models*. Lawrence Erlbaum Associates, Inc.

Huber, R., Tonini, G., & Cirelli, C. (2007). Exploratory behavior, cortical BDNF expression, and sleep homeostasis. *Sleep, 30*(2), 129–139.

Hung, D., Shu-Shing, L., & Lim, K. Y. (2012). Authenticity in learning for the twenty-first century: Bridging the formal and the informal. *Educational Technology Research and Development, 60*(6), 1071–1091.

Leonard, S. N., Fitzgerald, R. N., Kohlhagen, S., & Johnson, M. W. (2017). 02-01 Design principles as a bridge between contexts: From innovation in the science museum to transformation in formal education. *Educational Design Research, 1*, 1–15.

Lin, P. Y., & Schunn, C. D. (2016). The dimensions and impact of informal science learning experiences on middle schoolers' attitudes and abilities in science. *International Journal of Science Education, 38*(17), 2551–2572.

Meltzoff, A. N., Kuhl, P. K., Movellan, J., & Sejnowski, T. J. (2009). Foundations for a new science of learning. *Science, 325*, 284–288.

Murphy, C. (2015). *Reconceptualising the learning and teaching of scientific concepts.* In D. Corrigan, C. Buntting, J. Dillon, A. Jones, & R. Gunstone (Eds.), *The future in learning science: What's in it for the learner?* (pp. 127–150). Springer. Retrieved from https://doi.org/10.1007/978-3-319-16543-1

Murphy, C. (2016). *Coteaching in teacher education*. Critical Publishing Ltd.

Palincsar, A. S., Brown, A. L., & Campione, J. C. (1993). First-grade dialogues for knowledge acquisition and use. In *Contexts for learning: Sociocultural dynamics in children's development* (pp. 43–57). Oxford University Press.

Pelligrini, A. D., & Holmes, R. M. (2006). The role of recess in primary school. In D. Singer, R. Golinkoff, & K. Hirsh-Pasek (Eds.), *Play=learning: How play motivates and enhances children's cognitive and socio-emotional growth*. Oxford University Press.

Robbins, D. (2003). Vygotsky's zone of proximal development, imitation and mimesis: Thoughts regarding L2 and psycholinguistics. *Kul'turnoistoricheskii Podkhod i Problemy Tvorchestva*, 52–57.

Scott, P., & Mortimer, E. (2005). Meaning making in high school science classrooms: A framework for analysing meaning making interactions. In K. B. Martin, G. Onno de Jong, & H. Eijkelhof (Eds.), *Research and the quality of science education* (pp. 395–406). Springer.

Stocklmayer, S. M., Rennie, L. J., & Gilbert, J. K. (2010). The roles of the formal and informal sectors in the provision of effective science education. *Studies in Science Education, 46*(1), 1–44. Retrieved from https://doi.org/10.1080/03057260903562284

Vygotsky, L. S. (1978). *Mind in society: The development of higher psychological processes* (M. Cole, V. John- Steiner, S. Scribner, & E. Souberman, Eds.). Harvard University Press.

Vygotsky, L. S. (1986). *Thought and language*. MIT Press.

Vygotsky, L. S. (1997). *The collected works of L. S. Vygotsky, vol. 4. The history of the development of higher mental functions* (R. W. Reiber, Ed.; M. J. Hall, Trans.). Plenum.

Watson, J. D., & Crick, F. H. C. (1953). A structure for deoxyribose nucleic acid. *Nature, 171*, 737–738.

# Chapter 9
# Vygotsky and Science in Higher Education

## 9.1 Introduction

The teaching of science at third level in universities and colleges is very different from that of school, particularly regarding the practice of lecturing to large groups of students, routinely carried out in teaching first- and second-year science students. Traditional science education at third level is considered to be less appealing to science students, and accounts partially for increasing attrition rates, particularly in the physical sciences.

The aspirations of Project 2061, launched as long ago as 1986, by the American Association for the Advancement of Science (AAAS), aimed to improve general levels of scientific literacy by the year 2061. Its founder, F. James Rutherford, stated that:

> *The life-enhancing potential of science and technology cannot be realized unless the public in general comes to understand science, mathematics, and technology and to acquire scientific habits of mind; without a scientifically literate population, the outlook for a better world is not promising.* (Rutherford & Ahlgren, 1991, pp. v–vii)

Thirty years on, it is the undergraduate science students, not only the public, who have yet to acquire scientific habits of mind as part of their formal undergraduate education. Acquiring scientific habits of mind includes investigating phenomena in lab classes and applying explanations to solving realistic problems. Some examples of the scientific habits of mind include:

- Curiosity.
- Scepticism.
- Openness to new ideas.
- Creativity.
- Intellectual honesty.
- Ethical responsibility.



C. Murphy, *Vygotsky and Science Education*,
https://doi.org/10.1007/978-3-031-05244-6_9

Such habits are encouraged when Vygotskian ideas of social constructivism are integral to students' experiences of practical learning science, such as:

- Designing, modelling, and executing experiments that test hypotheses.
- Constructing knowledge from the outcomes of their experiments.

A few years ago, I worked with physics and mathematics colleagues to explore a model of problem-based cooperative learning (PBCL) in which one cohort of students was tasked in groups to determine experimentally a value for the acceleration due to gravity, whilst a second cohort followed a more traditional lab-manual approach, usually as individual or paired learners (Bergin et al., 2018). The students who followed the PBCL approach appeared more engaged, and their perceptions of the scientific process emphasised creativity and criticality. Students who followed the lab-manual approach thought the experiment was "easy".

This chapter focuses on three elements of Vygotskian theory which help towards providing a framework for pedagogic innovation in higher education science learning and teaching. They are social constructivism, the ZPD and concept development. The main focus is on undergraduate student science learning, but most of the ideas and examples can also be applied in master's and doctoral science programmes.

A key global problem is the relatively high attrition rates among science undergraduates and postgraduates from universities. There are many reasons for this, including student lack of engagement with a wide range of undergraduate science courses, the perception that much of the content is too difficult, and low performance in science examinations. Science educators have considered problems with the way undergraduate science is taught in universities and colleges for more than 100 years, decrying too much emphasis on didactic teaching via lectures and heavily structured laboratory manuals. In 1894, the physicist Thomas Preston suggested:

> *It cannot be too soon or too often impressed upon the beginner that an acquaintance with a number of facts does not constitute a scientific education … Knowledge is not the mere memory of facts, but the comprehension of their whole meaning in the story of nature.* (Preston, 1894, p. 25)

Forty years later, John Dewey asked:

> *Why is it, in spite of the fact that teaching by pouring in, learning by passive absorption, are universally condemned, that they are still so entrenched in practice?* (Dewey, 1916, p. 46)

Of course, there are factors relating to attrition which are external to universities that cannot be addressed by changes to science pedagogy, but there are ways in which pedagogy can be enhanced to develop students' science *habits of mind*, such as critical thinking and reflection, imagination and creativity, curiosity, scepticism, team-working, and problem-solving. Several interventions and innovations have been introduced and shown to improve science habits of mind, such as interactive response systems; for example: 'clickers'(wireless devices used to conduct student participation activities in the classroom), e-learning, the use of specific apps, wikis etc., inquiry-based learning, cooperative problem-based learning, and other forms of active learning.

## 9.2 Social Constructivism

Social constructivism is a term used frequently in this book. Basically, it draws on constructivism, emphasising the *collaborative* nature of learning. Vygotsky rejected the assumption that it was possible to separate learning from its social context, as his CHT suggests. He argued that all cognitive functions originate in, and are products of, social interactions. According to Vygotsky, language and culture play essential roles both in human intellectual development and in how humans perceive the world. Human language enables us to impose culturally defined sense and meaning on the world. Language and culture are the frameworks through which humans experience, communicate, and understand reality. One of Vygotsky's famous quotes states:

> *A special feature of human perception ... is the perception of real objects ... I do not see the world simply in colour and shape but also as a world with sense and meaning. I do not merely see something round and black with two hands; I see a clock and I can distinguish one hand from the other.* (Vygotsky, 1978, p. 39)

In science and other subjects that require specific terminology (e.g., music, social science) and conceptual schemes that are transmitted by means of language, are essentially social phenomena. Thus, Vygotsky maintained that concepts are socially constructed. Knowledge is not simply constructed: it is co-constructed.

### *9.2.1 Implications for Undergraduate Science Teaching*

One of the key researchers in changing undergraduate teaching and learning is Eric Mazur. Mazur is Balkinski Professor of Physics and applied physics at Harvard University. He is also a passionate advocate for improving undergraduate science teaching that involves social constructivist approaches, including the use of digital technology resources. His own teaching experience was based on sharing information, including sharing his lecture notes at the end of the class, which made him question the value of transmissive teaching.

Mazur began to question his own teaching approaches after observing his students' relatively poor improvement on the Force Concept Inventory (FCI) test, despite a semester's teaching. The FCI is designed to assess student understanding of the Newtonian concepts of force and is widely held as the 'gold standard' conceptual inventory in the physical sciences. As a consequence of the lack of improvement after traditional teaching, Mazur began to think of his teaching as more than the transfer of information. He created opportunities for students to digest that information and apply it within the realm of their own experience – and in contexts beyond their own, working collaboratively (Fagen et al., 2002).

A key idea was that the learning distance between most of the students and the expert (Mazur, in this case) was too great – outside their zones of proximal development (ZPDs). So, in lectures, he invited the class to discuss the problem with each other and within minutes found they had figured it out. Sharing the ideas with a few other students who had just learned about the concept to solve, a problem which might sound simple enabled them to pool their knowledge and come up with a solution.

Mazur utilised the interactive response technology of 'clickers' to generate a plot of their individual answers initially and after the collaborative discussions, noting the significant change that showed few correct solutions the first time and mostly correct the second time. The shift in learning was characterised by making the lesson more active (not passive), and allowing for the personalisation of learning, thus enabling students able to engage with each other in ways that an individual teacher never could. The focus of the exercise moves from knowing a fact to generating curiosity, thus motivating students to share knowledge as a more effective way to problem-solve.

Another key idea in Vygotsky's *oeuvre* is that of providing time for brain processing in concept development – the more social interaction in the process, the higher chance that students will be able to grasp the concept more fully. Mazur, in a keynote speech (summarised in a blog by Tessa Gray, 2017) referred to Vygotskian theories when discussing that education as a social experience, not an isolated one.

Mazur also used evidence for active versus passive learning from research carried out by Poh et al. (2010), in which subjects were fitted with wrist sensors that measured skin conductance as an index of the "arousal associated with emotion, cognition and attention" (p. 1243). Mazur presented a figure from the Picard group's paper showing wrist-sensor readings for a single MIT student over the course of week (see Fig. 9.1). The sensor recorded regular, strong spikes during periods of study, lab work and homework, but the readout flatlined during two activities: attending lectures and watching TV.

A PDF of the Poh, Swenson and Picard paper, "A Wearable Sensor for Unobtrusive, Long-Term Assessment of Electrodermal Activity," is available online.[1] This research clearly showed that for the student under study, attending science lectures showed almost no activity that related to emotional engagement, which, according to Vygotsky and others, is an absolute requirement for learning to occur.

Mazur's focus on social constructivism has led to his role in developing technologies which facilitate social learning. One such is Perusall,[2] a social learning platform developed by him and his colleagues, providing powerful asynchronous learning experiences for students out of class. Such software provides a rubric-based assessment approach, which can be used as a powerful way of engaging students and enabling them to take ownership of their learning. This social interaction provides the intrinsic motivation for learners to learn. Vygotsky maintained that true learning is that which is learning for the sake of learning, as opposed to learning for

[1] https://affect.media.mit.edu/pdfs/10.Poh-etal-TBME-EDA-tests.pdf

[2] https://perusall.com/

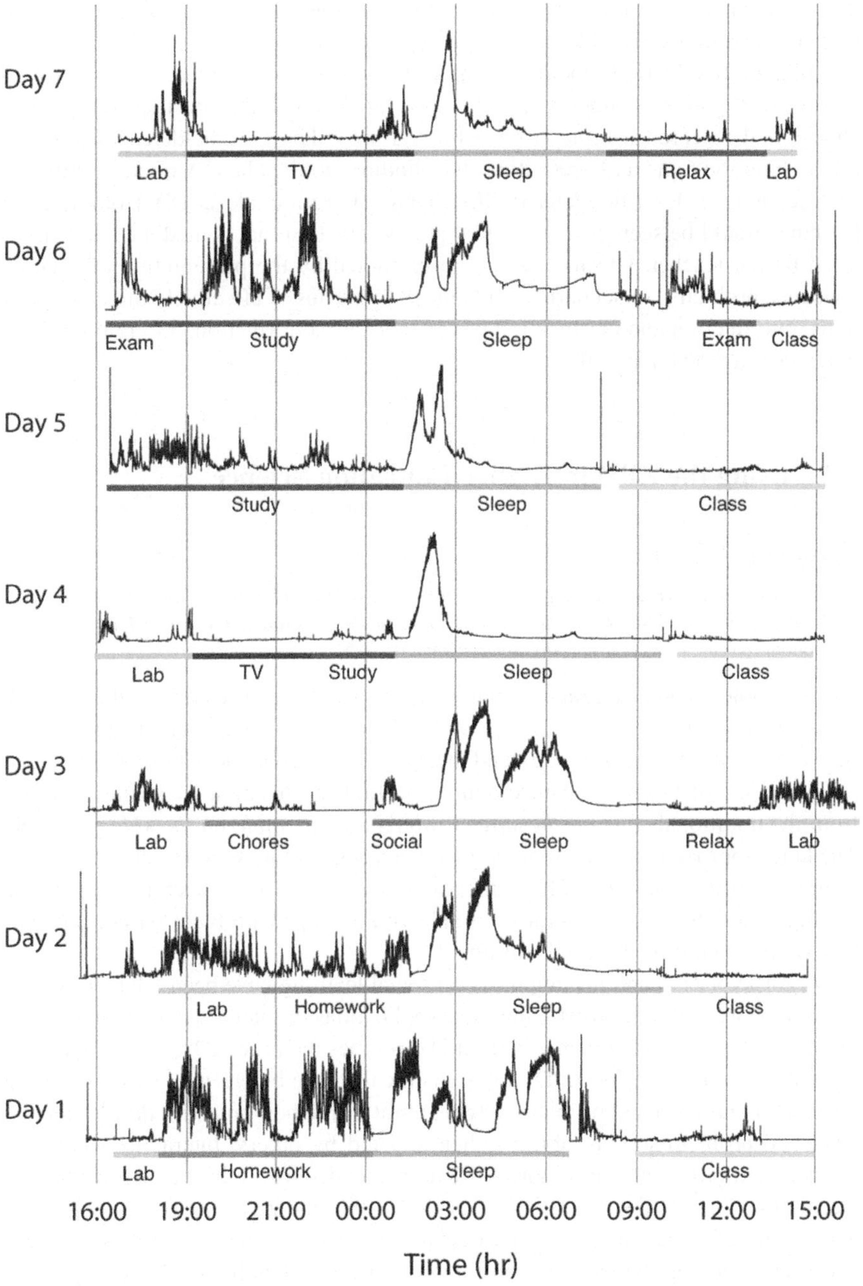


**Fig. 9.1** Wrist-sensor readings for a single MIT student over the course of a week

short-term gain, like studying for an examination. It is common for such learning to be transient and irretrievable not long after sitting an exam.

Collaborative learning methods require learners to develop teamwork skills. In group investigations, students may be split into groups that are then required to choose and research a topic from a limited area. They are held responsible for researching the topic and presenting their findings to the class. A good example of such learning is described later in this chapter (Bergin et al., 2018). Collaborative learning should be seen as a process of peer interaction that is mediated and structured by the teacher. Discussion can be promoted by the presentation of specific concepts, problems, or scenarios, and guided by means of effectively directed questions, introduction and clarification of concepts and information, and references to previously learned material.

## 9.3 Using the ZPD in Higher Education Science

Vygotsky (1997) argued:

> *That the school has been locked away and walled in as if by a tall fence from life itself has been its greatest failing. Education is just as meaningless outside the real world as is a fire without oxygen, or as is breathing in a vacuum.* (p. 345)

This quote could also relate to many undergraduate science courses that are still dissociated from real-world science and still lack the engagement factors which 'hook' students into a desire to extend their learning beyond the lecture. Fortunately, the pedagogy of teaching science courses in higher education is starting to move from the traditional *sage on the stage* towards a more *guide on the side* approach. The latter approach requires changing from a teaching paradigm towards a student-centred learning paradigm. Student-centred learning is more experienced-based. Examples abound, such as inquiry-based learning, flipped, problem-based, cooperative, digital, and project-based learning.

Vygotsky's ZPD and his emphasis on social learning have been adopted widely as both theoretical frameworks and their applications in science teaching and learning in universities and colleges. The ZPD describes the process by which individuals learn from others. The ZPD represents the distance between a learner's current developmental level and problem-solving ability, and their *potential* developmental level and problem-solving ability when assisted by others. Interactions between learners and others who are more skilled or knowledgeable in the task enable the learner to complete the task alone. Working within the ZPD is a two-way process and often leads to learning for all participants, particularly in their abilities to clarify and share meaning, which can lead to a deeper understanding of the topic or process. The combination of learning via discussions *with* practical activities are considered to contribute more towards cognitive development than by working alone.

### 9.3.1 Formation of the ZPD

Vygotsky suggested that an individual's actual level of development as determined by independent performance, such as an IQ test: "not only does not cover the whole picture of development, but very frequently encompasses only an insignificant part of it" (Vygotsky, 1998, p. 200).

He reported that responsiveness to mediation is required for understanding cognitive ability because it provides insight into the person's future development. Thus, what the individual can do one day with help, they can do tomorrow alone. Potential development varies independently of actual development, meaning that the latter cannot be used to predict the former. Vygotsky's work illustrated that test such as the IQ measure knowledge that is already known, but they give no indication of the learner's ability to build on or extend that knowledge.

Vygotsky considered performance on summative tests as an indication of the already known 'past' knowledge and argued that "instruction must be orientated towards the future, not the past" (Vygotsky, 1962, p. 104).

Applying this research to higher education science indicates that ways of assessing students need to be developed, such that low levels of recall (the bottom level of Bloom's taxonomy of knowledge) are not the key measure of attainment. There needs to be a move towards assessing higher and different forms of science knowledge and expertise at undergraduate level, particularly in the early years of study. And more use of continuous assessment which includes teamwork, giving presentations, developing position papers and other activities which allow for assessment of learners' potential, will help learners to navigate through the ZPD.

### 9.3.2 Key Elements of the ZPD for Developing Teaching

Learners' behaviour, as it affects learning, is the realisation that learning can be difficult, and that it does not always assume a smooth upward trajectory. Regression and recursion are key to deep learning. A most useful illustration of regression and recursion was developed by Tharp and Gallimore (1988), illustrated in Fig. 9.2.

An example of de-automatisation in science learning could be observed in relation to learning genetics. Students can be taught all the requisite terminology and the way to perform genetic crosses successfully, but after time and without repeated practice, the ability to perform such crosses needs to be developed anew. Further learning by an individual is made up of these same regulated ZPD sequences, from other- to self-assistance, recurring repeatedly for the development of current and new capacities.

There is a mix of other-regulation, self-regulation, and automatised processes for each learner at any one time. Therefore, even the expert can benefit from regulation for enhancement and maintenance of performance (e.g., teachers undergoing CPD). Gallimore and Tharp (1990, p. 187) further suggest that de-automatisation and

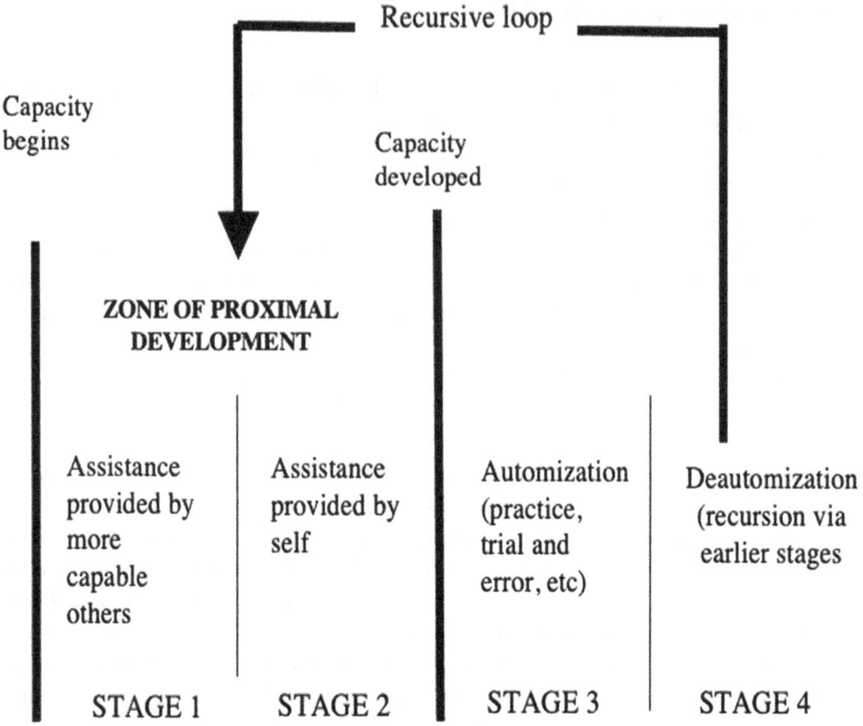


**Fig. 9.2** My adaptation of the ZPD by Tharp and Gallimore (1991)

recursion occur regularly: "What one formerly could do, one can no longer do". It often happens that self-regulation is not sufficient to restore performance capacity after de-automatisation and thus other assistance is again required.

Some reasons for de-automatisation, suggested by Dunphy and Dunphy (2003) in the context of surgical training, could be "environmental change, stress, major upheaval and trauma" (p. 50). In the case of school learning, one major reason for de-automatisation could be he variety of different experiences the student has undergone between successive learning situations. De-automatisation as a concept is most helpful in designing ZPDs which allow for cycles of assisted and non-assisted task completion in the development of scientific concepts.

### *9.3.3 Practical Examples of Using the ZPD in Undergraduate Physics Teaching*

A useful summary of the ZPD comes from Wass and Golding (2014, p. 671), which suggests that the basic idea is that "…we should pitch what we teach so that it is slightly too hard for students to do on their own, but simple enough for them to do with assistance". However, this idea could be contested on the grounds that it might assume a one-way relationship between 'learner' and 'teacher'. A social constructivist approach to science education research and practice positions learners and teachers *both* contributing to ZPD development and promotes for collaborative and active learning approaches.

The core idea is that students can operate at a higher and more independent level than they could on their own if they were provided with more capable peer or teacher assistance. The two key 'scaffolds' to facilitate development through the ZPD are *structuring the task* and *problematising* (Wass & Golding, 2014).

These ideas about the ZPD in science learning in higher education were useful in developing a pilot study to improve the learning experience of engineering students in their studies of a freshman year physics course (Bergin et al., 2018). They adopted a learner-centred instructional approach, PBCL, in which students, working as collaborators, took responsibility for their own learning as problem-solvers in their practical lab classes. Two cohorts of students (each comprising approximately 100 students) carried out an investigation on finding a value for acceleration due to gravity, one using the traditional lab-manual-led approach (control group), whilst the others (PBCL group) worked in small teams and were encouraged to create, construct and critique their own approaches.

The lab-manual-led cohort was asked to find a value for acceleration due to gravity using a pendulum, an equation, and an explicit step-by-step procedure given in the lab-manual. Students worked in pairs and, at the end of the three-hour laboratory session, reported on their findings in a pre-designed template. A student's work was assessed against their ability to follow procedure and arrive at a pre-determined solution.

For the PBCL approach, students worked in teams of three or four. They were posed with an experimental problem, and encouraged to construct and critique their own approaches, using any equipment or technology available to them in the lab or on-line. They were introduced to the lab (in groups of approximately 20 students) before starting, in small teams, to develop their strategies for solving the problem. A demonstrator was on hand for advice when required. After an hour, student teams presented their initial ideas, strategies, and findings to one another. They were encouraged to peer-review each other's strategies, suggest modifications to one another's strategies, and use ideas from the review session in their own team's experiments. Approaches employed to determine acceleration due to gravity varied greatly from those covered in standard university physics textbooks to more creative methodologies. PBCL students recorded their experimental findings in a 'free' format and were encouraged to record all of their ideas and thinking in attempting to find a value for acceleration due to gravity. Their reports were assessed in terms of their creativity, as well as the reliability and validity of findings. A similar completion rate of the submitted lab report was seen in both cohorts.

In terms of the ZPD, it was observed that members of the PBCL group were facilitated to develop their ideas in a much-expanded ZPD, which comprised peers for discussion, freedom to access information and potential strategies from a much broader source, including internet, videos, on-line communication, choice of materials where possible, excitement, teamwork, curiosity, thinking 'outside the box', freedom and challenge.

In Vygotskian terms, the difference in breadth of conceptualisation of the scientific process between the two groups could be explained by the narrow ZPD (development of skill manipulation) within which the traditional group of students were learning. The more expanded ZPD provided by the PBCL approach encouraging students to use and interrogate scientific principles in their design method and data interpretation using dialogue, discussion, experimentation with ideas, to fail, try again, persevere, trust, critique, create and discover, could be key to promoting bigger-picture thinking about science.

Analysis of the student perspectives indicated that responses from PBCL students indicated more engaged and broader thinking than those in the control group. Students who participated in a PBCL approach to labs reflected on enjoying the challenge of designing their own experiments and demonstrated a better understanding of the scientific process in devising and verifying their own experimental work via practices associated with constructing their knowledge of physics and scientific endeavour. Students' responses from the PBCL cohort contrasted with those of students in the control cohort, who showed an alternative understanding of the scientific process from their learning experiences in the lab (e.g., following procedure) and were not as positively engaged in their learning.

These findings are in line with advice from a recent report from the American Association of Physics Teachers (AAPT) on undergraduate laboratory curricula that emphasised 'sensemaking' strategies over excessively procedural approaches (Kozminski et al., 2014). Sense-making comes from the learner struggling, usually with others directly via collaboration, or more indirectly via books, computers, etc. – all of which have been written or programmed by others. Vygotsky referred to sense-making as the learner becoming the source of their own learning. Most traditional approaches, by which teachers structure the content and guide students through it, represent that in which *teachers are the source of learning*. Creating conditions which encourage students to ask questions, try out strategies, etc. as in PBCL is spontaneous learning, in which the *students are the source of their own learning*.

## 9.4 Concept Development in Higher Education Science

Davydov (1930–1998) was a prominent educationalist in the Vygotskian tradition, famous for his significant work on generalisation and the practice of inquiry-based teaching. This section introduces Vasily Davydov's (2008) conception of theoretical thinking in terms of concepts which illustrates the complexity of scientific concepts and their existence only in relation to other concepts:

> *...the basis of theoretical thinking, which operates not with conceptions but rather with proper concepts... a concept is the form of thinking activity that reproduces an idealized object together with its system of links. In their unity, these links reflect the universality or*

> *essence of the movement of the material object. The concept is simultaneously* ***both a form of reflection of the material object and the means for psychologically reproducing or constructing it.*** *That is, the concept is a special thinking action.* (pp. 90–91, emphasis added)

Concepts in science learning and teaching are often used as singularities, which exist on their own. Such practice leads students to develop a lower form of concept, described by Vygotsky as a "pseudo concept", which can be confused with true concepts because the learner might be using the right words to describe the concept, but lacking the logical connections between its parts. The learner can use the pseudoconcept in communication and activities, such as exams, as if it were a true concept. For example, the learner may use the definition of an electron transport chain in the context of photosynthesis metabolism but cannot relate it to other contexts.

The special *thinking action* of concepts, as noted by Davydov (2008), has its genesis in the development of scientific or abstract concepts in childhood as discussed by Vygotsky. Everyday and scientific concepts differ in terms of their history of development (Vygotsky, 1987). Everyday concepts are learned and developed through experiences. Scientific concepts are those that are taught and often experienced, at least initially, as abstractions. The richness of everyday concepts provides the basis for development of scientific concepts or abstracted forms of a concept. Current research highlights the rather common issue of holding potentially contradictory concepts; as one example, an everyday concept of 'fish' that contradicts the scientific concept of 'fish' (Karpov, 2003). A central concern for educators, then, is how to integrate everyday concepts and scientific concepts into a dialectically logical system (Howe, 1996).

In contrast to what is promoted in many learning contexts, Vygotsky (1987) argued that the process of concept formation is not just an act of generalisation on the part of the learner, but it considers both the everyday conception of a phenomenon and an abstracted reading of the same experience. Whilst everyday and scientific conceptions develop differently, they are always related to each other, and always united in a single, albeit at times contradictory, system of conceptual knowledge construction that is culturally formed communities or as part of professional and scientific knowledge construction. The emphasis here is not on the solitary learner, but on interacting, sharing, and negotiating meaning to integrate everyday concepts into a system of related concepts.

The complexity of integrating everyday and scientific concepts is a contributing factor in student engagement/disengagement in science subjects. Indeed, Howe (1996, p. 48) raised some important research questions based on a Vygotskian approach to science learning:

> *What problem solving strategies do children use in everyday life that have been ignored in school and can be used as a basis for science teaching? What are the differences between the everyday science concepts of children from different socioeconomic, ethnic and regional backgrounds and how does this affect what is learned?* (p. 48)

Scholars have long highlighted the potential incongruity between learning in and out of schools (Resnick, 1987; Sefton-Green, 2012), in addition to the differences in how everyday and scientific concepts develop, the role of teaching and learning in this development, and the often inauthentic and abstract ways in which the structure of undergraduate science courses prioritises curriculum delivery over deep and relevant study.

The vast literature on gender and science and mathematics related fields contributed initial momentum and a vision for science education that engages both females and males in preparation for science and mathematics careers as a method for overcoming the bias and male dominance in these fields. Democratisation of scientific institutions is an ideal, but many concerns remain (e.g., Bottia et al., 2015) and have been extended to the ways in which socioeconomic, ethnic, and language differences intersect to steer students away from science subjects.

The profound sociality of concepts, as well as teaching and learning, further emphasises the importance of authentic engagement in problems that matter to participants. As noted by Wells (2008), scientific concepts are not possessed by individuals; rather they are historical, social, and cultural resources that are used for a variety of purposes. Scientific concepts are 'cultural tools' developed by scientists, to help describe and explain the world around us. Mastering their use, Wells (2008) suggested, is best developed when students are engaged in scientific problem solving that requires these 'tools'. The concepts that develop emerge when the relationships between teachers, students, and curricula are brought together to attend and inquire into academic questions that are relevant to those involved.

The issue of relevance to learners is central for this discussion because it highlights *how* educators engage students in inquiry. Like the scholars and scientists before them, students arrive at school from multiple locations with different experiences and cultural expectations, along with their own needs, interests, and motives (Hedegaard & Chaiklin, 2005). Human development occurs in relation to access to shared collective memories, or knowledge, that have been recorded and passed on through writing. This shared knowledge was more permanent and, thus, independent of those who produced it, becoming a written 'objects' for education in different cultures (Wells, 2008).

Knowledge was passed on via collaborative activity using cultural tools, such as pictures, diagrams and writing. As science developed, concepts were created, based on empirical observation and thorough scientific investigation, to help explain phenomena. Yet, and consistent with Vygotsky's and Davidov's emphasis on teaching the history of ideas, the focus on teaching the 'complete' concept in schools often erases the complexity of knowledge construction: the history of ideas, the disagreements, the misunderstandings, and the power inherent in which and whose knowledge is valued.

From the narrative of Galileo's letter to the Grand Duchess Christina, to Thomas Kuhn's notion of 'popular science', to the complicated history of the theory of continental drift, scholars have argued for learners to become connoisseurs of science, which includes increased attention to discussions of how, where and when various scientific concepts came about, especially the associated difficulties, political and technological barriers and enablers, as well as other human factors, to engender a deeper appreciation of the scientific endeavour.

In his theoretical consideration of concepts, Vygotsky used a model from classical mathematics to suggests that, ultimately, concepts are all subsumed into one dialectical system that he referred to as a system of equivalences:

> *The higher levels in the development of word meaning are governed by the law of equivalence of concepts, according to which* ***any concept can be formulated in terms of other concepts in a countless number of ways.*** (Vygotsky, 1986, p. 199, emphasis in original)

His broad grid for concepts is based on the surface of a globe, onto which every concept can be placed using a system of coordinates, corresponding to latitude and longitude in geography. A concept's 'longitude' relates to its degree of abstraction, and thus characteristic of thought processes, whilst its 'latitude' represents its objective reference, for example: plant or animal (Vygotsky, 1986, pp. 199–200).

However, the geographic analogy is only useful at a surface level. Vygotsky himself emphasised the limitation of the geographic analogy as being neither complete nor accurate, although it has been used since, particularly in philosophical considerations of concepts, such as in the work of the 'ordinary language' philosopher, Gilbert Ryle. Vygotsky contended that in a *true scientific concept*, the bonds between the parts of an idea and between different ideas are dialectically related; thus, the ideas form part of a socially constructed and accepted system of hierarchical knowledge (Roth & Lee, 2007). What was also obvious to Vygotsky was the necessity of change over time – change in concepts and knowledges in relation to changing conditions – as well as both the recognition of developments in science leading to new knowledge and the ease with which this knowledge can become coopted by political and economic interests and motives. It was for these reasons that he argued that freedom is the ability to break from current conditions, to imagine beyond what exists to what could exist (Vygotsky et al., 1994).

In teaching, Vygotsky proposed that teachers create a ZPD between the scientific and everyday concepts by illustrating and emphasising the *relationships* between them and showing how the scientific concept can be utilised to explain the everyday concept, whilst simultaneously raising the everyday concept towards its scientific conceptualisation (see Chap. 3, Fig. 3.1). For instance, a child may have a rich understanding of the everyday concept *brother* but not be able to define it in the more logical, conceptual way as *male sibling* (Panofsky et al., 1990). The task of the teacher, for Vygotsky, is not to evaluate individual conceptions as correct or as 'misconceptions', but rather to help the child, through instruction with respect to the *relationship* between concepts within a system of concepts, and to develop conscious awareness and voluntary control of their own thinking (Wells, 2008).

Vygotsky (1987) argued that a "concept must be seen as part of the entire system of the relationships … just as a stitch must be seen as part of the fibres that tie it to the common fabric" (p. 193). This underpins theoretical knowledge and thinking (Davydov, 2008). The system is captured through the relations between everyday concepts and scientific concepts embedded in meaningful practices of a particular community that a child is oriented towards motive orientation (Hedegaard & Chaiklin, 2005). When communities foreground the learning of scientific concepts in support of a motive orientation leads to more democratic and equitable outcomes for diverse students.

Vygotsky's fascination with teaching and learning, integrating everyday and scientific concepts, and contributing to social change was key to his idea that society needs to provide tools to engage learners with different characteristics and abilities. While engaging diverse learners and enabling them to access knowledge was a significant intergenerational responsibility and achievement, the purpose of doing this was to support the development of cultural tools that could be taken up, used, and transformed in their application to current and future concerns. It was expected that human action would transform the tools with which they think, create, and transform their worlds. Science continually puts to the test ways of thinking and acting and, throughout this process, opportunities arise to ensure that the direction of change in science and science education is towards becoming more equitable and inclusive. Vygotsky's equity orientation drives this process.

## 9.5 Summary and Conclusion

This chapter has considered a variety of sources of evidence for the recent increase in popularity of Vygotsky's research on the ZPD and concept development, together with the overall adoption of more social constructivist approached to teaching and learning in higher education.

The challenges of science teaching in large lecture theatres, despite the support of digital learning resources which are designed to promote more collaboration, are being addressed by scholars who are disappointed in student exam performance. It has also discussed the increasing rates of student attrition. Eric Mazur's work has been highlighted, since it stems from his own experiences in the areas just mentioned. His work has contributed to the recent development of teaching in many universities to promote twenty-first-century learning. The key Vygotskian concepts used in this regard are social constructivism, the ZPD and concept development. Change is slow, but it has started. Examples of successful integration of Vygotskian constructs in higher education described in this chapter should provide some inspiration and practical guidance for science educators in third-level institutions.

The next chapter features a Vygotskian perspective of science teacher education, both in pre-service science teacher education, and CPD for in-service teachers. In addition, those who work with students in informal science contexts may also be interested in considering Vygotskian approaches to science teacher development in the next chapter.

## References


Bergin, S. D., Murphy, C., & Shuilleabhain, A. N. (2018). Exploring problem-based cooperative learning in undergraduate physics labs: Student perspectives. *European Journal of Physics, 39*(2), 025703.

Bottia, M. C., Stearns, E., Mickelson, R. A., Moller, S., & Valentino, L. (2015). Growing the roots of STEM majors: Female math and science high school faculty and the participation of students in STEM. *Economics of Education Review, 45*, 14–27.

Davydov, V. V. (2008). *Problems of developmental instruction. A theoretical and experimental psychological study*. Nova Science Publishers, Inc.

Dewey, J. (1916). *Democracy and education*. Macmillan.

Dunphy, B. C., & Dunphy, S. L. (2003). Assisted performance and the zone of proximal development. *Journal of Educational and Developmental Psychology, 3*, 48–58.

Fagen, A. P., Crouch, C. H., & Mazur, E. (2002). Peer instruction: Results from a range of classrooms. *The physics Teacher, 40*(4), 206–209.

Gallimore, R., & Tharp, R. (1990). *Teaching mind in society: Teaching, schooling, and literate discourse*. Vygotsky and education. Instructional implications and applications of socio-historical psychology. In L. Moll (Ed.), *Vygotsky and education* (pp.175–205). Cambridge University Press.

Gray, T. (2017). *Eric Mazur Keynote: Innovating education to educate innovators*. Retrieved from https://www.youtube.com/watch?v=DoibiRIDZdw

Hedegaard, M., & Chaiklin, S. (2005). *Radical-local teaching and learning. A cultural-historical approach*. Aarhus University Press.

Howe, A. (1996). Development on science concepts with a Vygotskian framework. *Science Education, 80*(1), 35–51.

Karpov, Y. V. (2003). Vygotsky's doctrine of scientific concepts: Its role for contemporary education. In A. Kozulin, B. Gindis, V. S. Ageyev, & S. M. Miller (Eds.), *Vygotsky's educational theory in cultural context* (pp. 65–82). Cambridge University Press.

Kozminski, J., Beverly, N., & Lewandowski, H. (2014). *Recommendations for the undergraduate physics laboratory curriculum*. American Association of Physics Teachers (AAPT).

Panofsky, C., John-Steiner, V., & Blackwell, P. (1990). The development of scientific concepts and discourse. In L. C. Moll (Ed.), *Vygotsky and education: Instructional applications of sociohistorical psychology* (pp. 251–268). Cambridge University Press.

Poh, M. Z., Swenson, N. C., & Picard, R. W. (2010). A wearable sensor for unobtrusive, long-term assessment of electrodermal activity. *IEEE Transactions on Biomedical Engineering, 57*(5), 1243–1252.

Preston, T. (1894). *The theory of heat* (4th ed.). Macmillan.

Resnick, L. B. (1987). The 1987 presidential address: Learning in school and out. *Educational Researcher, 16*(9), 13–20+54.

Roth, W.-M., & Lee, Y.-J. (2007). "Vygotsky's neglected legacy": Cultural-historical activity theory. *Review of Educational Research, 77*(2), 186–232.

Rutherford, F. J., & Ahlgren, A. (1991). *Science for all Americans*. Oxford University Press.

Sefton-Green, J. (2012). *Learning at not-school: A review of study, theory, and advocacy for education in non-formal settings*. MIT Press.

Tharp, R. G., & Gallimore, R. (1988). *Rousing minds to life: Teaching, learning, and schooling in social context*. Cambridge University Press.

Tharp, R. G., & Gallimore, R. (1991). *Rousing minds to life: Teaching, learning, and schooling in social context*. Cambridge University Press.

Vygotsky, L. S. (1962). *Thought and language*. MIT.

Vygotsky, L. S. (1978). *Mind in society: The development of higher psychological processes* (M. Cole, V. John- Steiner, S. Scribner, & E. Souberman, Eds.). Harvard University Press.

Vygotsky, L. S. (1986). *Thought and language*. MIT Press.

Vygotsky, L. S. (1987). Problems of general psychology. In *The collected works of L.S. Vygotsky, vol. 1*. (pp. 243–285) (R. W. Rieber & A. S. Carton, Eds.; N. Minick, Trans.). Plenum Press.

Vygotsky, L. S. (1997). *The collected works of L. S. Vygotsky, vol. 4. The history of the development of higher mental functions* (R. W. Reiber, Ed.; M. J. Hall, Trans.). Plenum.

Vygotsky, L. S. (1998). The crisis at age seven. In *The collected works of L. S. Vygotsky, vol. 5* (pp. 289–296) (R. W. Reiber, Ed.; M. J. Hall, Trans.). Plenum.

Vygotsky, L. S., van der Veer, R. E., Valsiner, J. E., & Prout, T. T. (1994). *The Vygotsky reader*. Basil Blackwell.

Wass, R., & Golding, C. (2014). Sharpening a tool for teaching: The zone pf proximal development. *Teaching in Higher Education, 19*(6), 671–684.

Wells, G. (2008). Learning to use science concepts. *Cultural Studies of Science Education, 3*(2), 329–350.

# Chapter 10
# Vygotsky and Science Teacher Education

## 10.1 Introduction

*Science is built up with facts, as a house is with stones. But a collection of facts is no more a science than a heap of stones is a house* (Henri Poincaré, (1854–1912)[1]

Henri Poincaré was a mathematician, scientist, and philosopher, who drew attention to the fact that most people were incapable of understanding mathematics and science – more than 100 years ago. Poincaré called for teachers of science and mathematics to prioritise the issue. In a similar vein, John Dewey (1859–1952), the American philosopher, psychologist and educational reformer whose ideas have been influential in education and social reform, asked:

*Why is it, in spite of the fact that teaching by pouring in, learning by passive absorption, are universally condemned, that they are still so entrenched in practice?* (John Dewey, 1916 p. 46)

Most people still find mathematics and science difficult. So, teachers of science, technology, engineering and mathematics (STEM subjects) have a bigger challenge than those of other subjects, in their efforts to help students to achieve success in these areas. The key question is *how* to address the problem, especially in science teacher education. This chapter considers both initial and in-service science teacher education with an emphasis on Vygotskian ideas and practices and includes primary (elementary) and secondary-level teaching of science.

### *10.1.1 Science Teacher Education in a Historical Context*

We can think of science as culture, in a similar way to art, music and drama cultural worlds. Each of these cultural worlds has developed their specific terminologies and practices, which enables those involved to communicate and understand the cultural

[1]This quote is from La Science et Hypothèse (1908), 168. In George Bruce Halsted (trans.). Science and Hypothesis (1905), 101.



C. Murphy, *Vygotsky and Science Education*,
https://doi.org/10.1007/978-3-031-05244-6_10

world(s) in which they engage. These cultures are included, to some extent, in formal and informal education contexts, and are thus influenced by the culture(s) of the society at the time. Table 10.1 below illustrates some of the major shifts in societal demands and their science-related responses during the 20th and 21st centuries, from the 1950s (following the First and Second World Wars.

**Table 10.1** Major shifts in societal demands and their science-related responses (adapted from Kidman and Fensham (2020)

| 1950s–1970s | **Societal demands** | **Science education response** |
|---|---|---|
| | Cold war, sputnik flight (1959), DNA structure identified (1953), thus a need for more and better scientists. | Massive need for physics and chemistry to be taught to a high level, with much more laboratory exercises, following the 'scientific method'; aimed mostly at students in grammar and private schools. |
| 1980s | **Demand for a science for all** | **Science education response** |
| | Change in science-based technologies, impacting positively and negatively on the health of the environment. Need for all citizens to keep up with these developments. | School curricula were changed to encourage full participation in a technological society, aimed at *all levels of schooling*. Focus on technology, as well as biology, chemistry and physics. |
| 1990s | **Politico-societal demand for choice in education** | **Science education response** |
| | Science for All needed to be operationalised, so national curricula and league tables were introduced globally; science and technology included as statutory subjects. Scientific literacy now had educo-political meaning, as with literacy and numeracy. Late 1990s – *Beyond 2000* report published to address the unobtained aims of 1990s' failures. | Nature of Science (NoS) was introduced into school curricula. Schools were provided with details of what and how to teach in science and technology. More emphasis on scientific inquiry in teaching. More emphasis on the narrative form in science pedagogy. Mandatory science for all students and optional courses for high achievers. |
| 2000s | **Societal demand for twenty-first century workforce** | **Science education response** |
| | PISA project started in UK – Assessment of 15-year-old students worldwide on cognition and scientific problem-solving. Later in the decade, the STEM agenda emerged to prepare students for the twenty-first-century workforce via 'STEM skills' | STEM activities introduced into schools to increase student interest in further studies in the STEM fields. Key focus on digital age, including the rise of social media. |
| 2010s | **Demand for change in STEM learning** | **Science education response** |
| | More emphasis on global problems; need for more collaborative science research and practice between countries, for example, climate change education, pandemic management. | Calls for twenty-first-century classroom to demonstrate:<br>1. Student-centred learning.<br>2. That education to be collaborative.<br>3. That learning should have context.<br>4. That schools should be integrated with society. |

The changes above indicate how science teaching and learning have been influenced strongly by societal demand. However, many of the ideals for effective science learning and teaching have not yet been reached for many reasons, including the lack of resources in schools, especially in relation to the time required for science teacher development (in and out of school), as well as more time for science teaching within an already crowded timetable. Many science teaching resources in schools are outdated and do not effectively support the science learning and teaching for the current societal demands, which include:

- Climate change education.
- Increasing marketisation.
- The rapid but uneven influence of information and communication technologies on the nature of learning and teaching.
- Shifts in the learning needs of students from literacy, numeracy, and content mastery to include soft skills like communication, curiosity, resilience, cooperation, and problem-solving abilities.
- The interests of many stakeholders in defining goals of education.

In 2020, the societal demands changed dramatically, due to the COVID-19 global pandemic, which caused school closures, amongst other restrictions required by lockdown processes aimed at reducing transmission of the virus. This brought about an increasing emphasis on digital learning and teaching. Unfortunately, this was hampered by the lack of digital hardware and in some regions, lack of internet. However, the increased use of technology in science learning has resulted in an expedited increase in digital learning, which was advocated well before the pandemic. The societal demands at the time of the pandemic resulted in an increasing recognition of the need for high-quality teacher education to improve the standard of education. These demands were highlighted even more strongly as society recognised the global challenges which need urgent attention.

The traditional conception of the scientific method was formerly described using five descriptors: empirical, replicable, provisional, objective, and systematic. In the twenty-first century, the scientific 'tools' have changed, for example: genome editing (e.g., the Crispr-Cas9 system), big data analysis, machine learning and artificial intelligence. The latter three tools are computer-based rather than lab-based, which is leading to new demands in the required skill sets for science students. The new tools also require increased awareness of data-related concepts, such as statistical analysis, from wider stakeholders including policymakers and the public. A good example of this has been the explosion of medical data generation and interpretation that has directly informed government interventions that affected the lives of billions of people during the COVID-19 pandemic that started in 2019/2020.

Scientific methodology has started the process of adapting to embrace such tools more fully, although much more effort is required to develop the teaching of scientific method which is more relevant to scientific research outside school. In addition to the tools themselves, the way science is enacted in the twenty-first century demands global collaboration between science and social science to address major environmental and healthcare challenges. For example, UNESCO has developed the Climate Change Education for Sustainable Development programme. This aims to help people understand the many impacts of climate change and to increase climate awareness among young people by encouraging innovative teaching approaches. Examples include integrating climate change education into existing school curricula, as well as enhancing non-formal education programmes through the media, and via horizontal networking and partnerships.

This chapter discusses Vygotskian theory and science teacher education, aimed at initial and in-service teacher education. Science teacher education is constantly under revision as educators, researchers, and policymakers seek to identify optimal models for learning to teach science. Some of the major issues with the learning and teaching of science in schools include the fact that:

- Most teachers are not specialised in *all areas* of science.
- Some students' lack of motivation to learn.
- Many students and teachers find abstract scientific and mathematical concepts difficult to grasp.
- School science does not represent much of current scientific research very accurately.
- School scientific resources (human and material) are frequently inadequate.
- Teacher shortages in many scientific subjects.
- There is a lack of in-service support for science teaching.
- There is a content-driven curriculum, which does not provide the capacity for investigative science learning and teaching.
- Many teachers of younger children lack confidence in science teaching.

Vygotskian theory and practice in science teacher education, both for initial and in-service science teachers, is discussed in this chapter. Some of the major contributions of Vygotsky's works to the preparation of new science teachers and science teacher further development include:

- Scientific concept development.
- The ZPD.
- The unity of affect and intellect (nothing can be learned without emotional engagement of some sort).
- Regression, recursion, and reflection.

## 10.2 College-Based Science Initial Teacher Education (ITE)

Most science ITE programmes take place in both college and schools. The college-based elements focus mainly on facilitating the pre-service teachers' development of pedagogical content knowledge (PCK), a useful construct developed by Shulman (1986) which refers to the specific knowledge teachers require to transform their subject expertise into facilitating student science learning. Figure 10.1 below illustrates this transformation in terms of linking theory and practice.

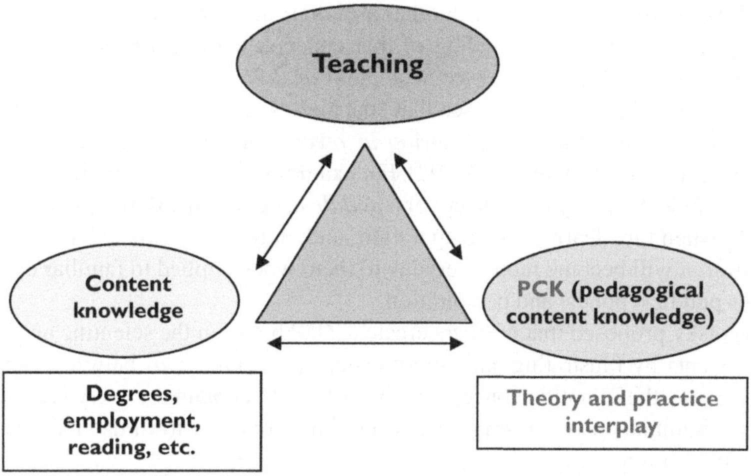


**Fig. 10.1** PCK as theory/practice interplay

Vygotskian theory can be most useful in developing PCK. His work provides a theoretical framework for helping students develop abstract scientific concepts, such as *evaporation* or the *ecological niche*. To be able to use such concepts successfully, students are required to raise their level of thinking. Vygotsky provides a useful mechanism for this. Another Vygotskian contribution, his distinction between the meaning and sense of words, is invaluable for pre-service science teachers to help students with the many difficulties that come from the use of specialised and sometimes highly complex scientific terminology.

### *10.2.1 Scientific Concept Development*

The focus on concept development in Chap. 7 (Sects. 7.5 and 7.6) provides detail of the process in secondary-level school science. The following, shorter account, gleans the key ideas from this work which are most relevant for pre-service teachers of science.

The key message in terms of the learning and teaching of scientific concepts is that they are viewed as dynamic, changeable, contextualised and usable *tools,* which have been developed over time to help explain the world and how it 'works' (Murphy, 2015).

Pre-service teachers (PSTs) can consider a variety of concepts that they might be using in their teaching and discuss with peers how useful they might be in terms of students' learning of science in the bigger picture. By doing so, they are using these concepts in the teaching context – some might be better than others in facilitating student understanding. Vygotsky argued that *all* concepts have both everyday (concrete, spontaneous) and scientific (abstract) dimensions. The teacher's job is to present both, as they relate to each other as a dialectical interaction. As the student develops a scientific understanding of that concept, their everyday idea becomes more scientific, and vice versa (see Sect. 3.3, Fig. 3.2).

Vygotsky (1934/1987) proposed that "the child's scientific and [her or] his spontaneous [everyday] concepts… *develop in reverse directions* (original italics) … they move to meet each other" (p. 192). For example: using the example of *evaporation*, the students' everyday concept of a *puddle disappearing* develops more scientifically when they learn about evaporation; and, at the same time, their concept of *evaporation* will become more everyday to them when applied to familiar contexts, such as puddles, ponds, and perspiration.

Vygotsky proposed that teachers create a ZDP between the scientific and everyday concepts by illustrating and emphasising the *relationships* between them and showing how the scientific concept can be utilised to explain the everyday concept, whilst simultaneously raising the everyday concept towards its scientific conceptualisation.

In considering how students develop scientific concepts, it is important that pre-service teachers can distinguish between the pseudoconcept and the true scientific concept. The pseudoconcept is knowledge of a definition, used by the student to answer an exam question which asks for that definition, but the student may have very little understanding of what it is in relation to other concepts. For example, the learner may use the definition of an ionic bond to describe how it differs from a covalent bond, without understanding the nature of chemical bonding. The words of the learner and teacher may refer to the same idea, but their meanings may not be the same (Gredler & Claytor Sheilds, 2008). A true concept is bound by *logical* bonds within parts and between different concepts.

The confusion between identifying pseudoconcepts and 'true'concepts accounts for the common experience of pre-service science teachers (PSTs) in that they understand science better when they start to teach it. They might have used personally meaningful pseudoconcepts to communicate knowledge successfully using the written form, including appropriate use of signs, symbols, and scientific terminology. But this may not have been as useful when trying to explain a similar idea without the 'props' of the signs and symbols.

True concepts are learned with conscious awareness and promote the development of everyday concepts into the accepted scientific framework, where they can be used, further developed and critiqued.

A major problem in concept development is recognition of when true conceptual thinking is being demonstrated. Unless this process involves evidence that learners are *using* the concept(s) appropriately, it could be a pseudoconcept, and not a true concept. Another issue is the case that learners can be thrust into problem-solving with new concepts before they have developed them sufficiently for the task, resulting in incomplete concept formation. The eventual formation of true concepts indicates that the learner is now able to master their own thinking. One of the most difficult tasks for learners is to learn the connections and relationships between concepts. A good idea is for them to construct a large visual diagram of the concepts in the topic, and between topics, as the term progresses. This activity requires pre-planning by the teacher to identify the required concepts for learning in advance.

### *10.2.2 The Zone of Proximal Development (ZPD) in College-Based ITE*

Vygotsky described the ZPD as a 'zone' in which a complex array of interactions between people and with their environment, serve to facilitate learners to reach the potential within a particular level of development. The ZPD could apply both to student learning in the classroom and pre-service teacher (PST) development as new teachers. Vygotsky stated that learning *leads* development (unlike Piaget, whose research led him to argue that learning follows development). Thus, Vygotsky described the 'internal' ZPD as psychological functions which are already maturing, and which can be nurtured to reach their potential with the help of more capable others, for example teachers, peers, teacher educators (see Fig. 10.2).

Vygotsky intimated that the actual development level characterises the cognitive development *retrospectively* – what they already know/can do, whilst the potential development level characterises it *prospectively* – what they are capable of with assistance. And he suggested that the best learning is that which takes place within the ZPD.

Pre-service teachers should consider the ZPD as one of the most useful constructs for their own, as well as for the students' science learning. The notion of creating ZPDs in their lesson plans to promote student learning (as opposed to focusing mainly on the resources) is very useful, particularly in terms of in-class collaboration within different science activities.

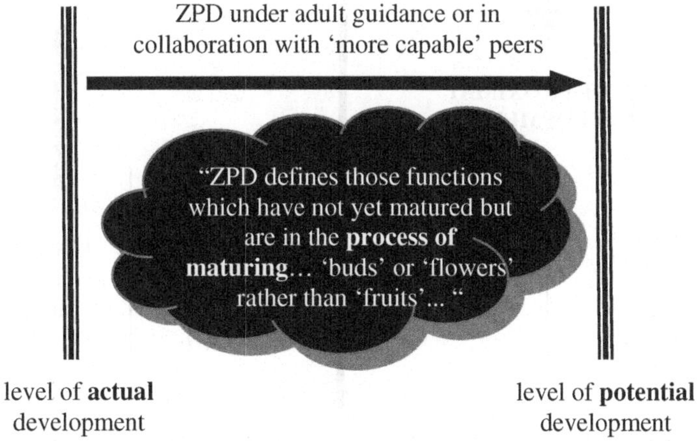


**Fig. 10.2** A visualisation of the ZPD as the 'zone of interactions' which facilitate learners in developing their potential

Another aspect of the ZPD is very useful, particularly for the development of skills. It was first mooted by Tharp and Gallimore (1988), who illustrated a four-phase ZPD (see Fig. 10.3 below). Their model addressed the development of any performance capacity based on the relationship between self-control and social control in an activity, which includes a 'recursive loop' in which learners revert to an earlier stage and progress through subsequent stages back to where they were – in effect they re-learn.

- In Stage 1 of Tharp and Gallimore's (1988) model, the learner relies on assistance (mostly via language) from more capable others to carry out a task.
- In Stage 2, the learner can self-assist.
- By Stage 3, the task is performed automatically.
- Stage 4 comprises the 'recursive loop' in which de-automatisation of performance leads to recursion back through the ZPD.

Examples of de-automatisation in teacher learning could be observed when a teacher is trying a new pedagogical approach, such as coteaching, or introducing new 'tools' (e.g., SMART Boards). In these situations, some 'automatic' aspects of teaching are lost, and need to be re-learned. Further learning is made up of these same regulated ZPD sequences, from other- to self-regulation, recurring repeatedly for the development of current and new capacities. There is a mix of other-regulation, self-regulation, and automatised processes for each learner at any one time. Therefore,

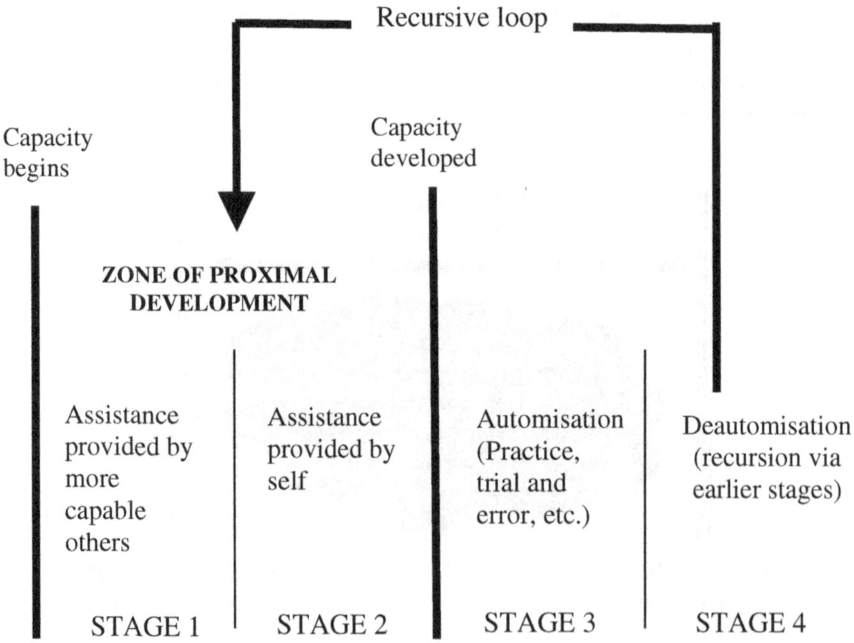


**Fig. 10.3** My Adaptation of Tharp and Gallimore's four-phase ZPD model (1998)

even the expert can benefit from regulation for enhancement and maintenance of performance (e.g., teachers undergoing professional development). Tharp and Gallimore (1988) suggested that de-automatisation and recursion occur regularly: "What one formerly could do, one can no longer do" (p. 187). It is frequently observed, however, that self-regulation is not sufficient to restore performance capacity after de-automatisation and so other assistance is again required.

Vygotskian reflection embraces consideration of situations from different angles. For example, children might be asked to write the Cinderella story as if she is the 'bad' character and her sisters are the 'good' ones. By doing so, the children's understanding of the original story can be assessed. Below is a sample scenario which can be 'performed' by pre-service teachers during a session in college, which takes place after they have already started their school-based teaching. After the performance, via guided reflection, they can work in groups to reflect on this vignette (below) from the supervisor perspective:

> A pre-service teacher, let's call her Sue, is about to start teaching a science class, as a supervisor walks in. Sue feels her face turning red, and her hands shaking.

(T) "Let's play a game!", exclaimed Sue, feigning confidence.

"When I call out a word/phrase, you call out whatever word or phrase you think in response. I'll write your words on the board. Just don't call until someone else has finished. OK! Here's the first: "being healthy".

(S) "Eating proper food"; "exercise"; "getting enough sleep"; "not smoking and drinking"

These responses came very quickly – Sue was relieved that they could play the game properly…

(T) "Good! Next: mobile".

(S) "Mobile phones"; "pay as you go"; "over babies' cots"; "being able to move about all right"; "Messi".[2]

A ripple of laughter moved across the room…

*(T)* "Messi? How did we get him?" Sue laughed and then all children laughed…

*(T)* "OK! Settle down! Here's another word: accessibility"

*(S)* "Credit card"; The class continued to laugh "hole in the wall"; "wall paper"; "paper underwear". More laughter and some inappropriate gestures…

*(T)* "Just a minute! These ideas are getting lost."

*(S)* "Lost? Missed penalty", shouted the boy who had first mentioned Messi – he stood up and started kicking balls of paper to a boy at the back of the room.

The responses came too fast for Sue to stop them. The Messi shout got an even bigger laugh this time – Sue realised she had lost the class.

*(T)* "All right, close your books and take out a pen" – she passed out the worksheet that she had planned as a group, open book activity: "You have 20 minutes to finish this test!"

(S) "You didn't tell us we were having a test!" "That's not fair." "We haven't even covered this yet." "I didn't do anything wrong." There were moans and looks of

[2] Messi is one of the best football players in the world at the time of writing.

disgust, even from the quiet students. "I'm reporting you to the principal; we have our rights!!!"

This comment hit home – the class had been studying human rights in the previous lesson. Sue felt terrible. How could she mark these tests? The first section had material they hadn't covered and the second was to work in groups to create a news-style programme

(T) "All right, all right – it won't be a test. But you do have to complete this for a coursework mark. You have proved that you cannot work together in groups today, so if your complete Part 1, we may have group work next lesson.

Sue was afraid to look at her supervisor. What was he writing down??

Pre-service teachers were then tasked to write if *they* were the supervisor? What advice might they give to the pre-service teacher? What learning took place in this scenario, and how can it be improved? What can be learn from the dramatic collisions Sue experienced? How did affect and intellect interact during this vignette? What lessons can be learned from this situation in terms of early practice in teaching?

This activity is an example of ZPD creation. Pre-service teachers were given the opportunity to raise their thinking level by imagining the assessor's views as well as their own, resulting in an expanded cognitive framework.

The idea of dramatic collision in learning is helpful in that in any learning situation, as only certain aspects lead to the creation of an 'aha' moment of understanding, or the need to find out more about something. Dramatic collisions in teaching can spark very effective learning, from interactions that cause both positive and negative emotion. For example, a classroom teacher reflecting on her own practice whilst observing the pre-service teacher:

> *... there were sometimes children in the classroom continually getting the attention from the student teacher because they were the loudest who were always coming up with answers, always being funny... and there were other children who were being completely ignored... because they were quiet and sitting not making a sound but not showing any interest. It made me aware that I'm probably doing that in my own teaching ...* (Primary classroom teacher [Roth & Tobin, 2006, p. 210)

Other researchers have indicated the importance of regression and recursion in development, for example, the image of ocean wave movement in relation to Vygotsky's work on development (it moves forwards and backwards, but always progresses in the appropriate direction). Using this metaphor, development is both progressive and regressive. When it is progressive, the wave becomes deeper and higher as it moves forward, which exemplifies the cumulative effect of increased development (Zebroski, 1994).

Early experimentation on the ZPD by Vygotsky involved problem-solving activities by children who all had the same score in IQ tests. This score was used by Vygotsky as representing the *actual level* of development. The children were then invited to solve a new problem, slightly more difficult than those on the IQ test, represented as the *potential level*. They received different types of support in solving the new problem, such as the teacher:

- Demonstrated how to solve the new problem and see whether the children could do it.
- Began to solve it and see whether the children could finish it.
- Asked the child to solve with the help of another child with a higher IQ.
- Asked leading questions.

They found many differences between the children's ability to solve the new problem, despite all having the same IQ score. Some children were not able to solve the new problem, despite the support offered. Thus – it could be argued that the IQ score is not a good predictor of a child's potential to learn. A better predictor is to look at the child's *potential* level of development, as opposed to the *actual*. The 'size' of the ZPD can be measured in terms of the amount of support required to solve new problems. However, the ZPD size is not fixed, and can depend upon the type, not the amount of support. The ZPD is a much better predictor than the IQ score of future intellectual development. This is because the ZPD directs attention towards teaching *already maturing* cognitive functions, not those which have already matured, or not even started.

The ZPD is a complex notion, which can be summarised as follows:

- It represents "the social origins of the whole process of children's cultural development and its relation to educational practice" Levykh (2008, p. 89).
- It is produced because of the difference between the level of natural, unmediated development and the level of development led by learning, mediated via language, artefacts, signs, gestures, and emotions.
- It is future oriented (past and current cognitive processes in the light of what is to be developed soon).
- Its size refers to the extent to which a learner can take advantage of collaborative interactions to realise future performance (Chaiklin, 2003).
- It is a "dynamic process that also reflects constant changes in the emotional connections among all participants" (Levykh, 2008, p. 91)
- Functionally, it is a complex, creative collaboration with others and through the environment, which together lead to new cognitive functions at a higher level of development.
- It is not only a process, but also a *synthesis of intellectual and emotional functions.*
- It encompasses both affective and intellectual features reflected in behavioural mastery (the highest form of human will power (Levykh, 2008). Behavioural mastery and development of self are interrelated, and both are mediated by emotions.

Finally, the ZPD could be considered as a complex interaction between participants, their collaboration, the types of 'tools' used, the type of mediation via the tools and the cultural contexts. Levykh (2008 p. 100) describes it as a "system of systems".

## 10.3 School-Based Science Initial Teacher Education (ITE)

Despite the challenges involved in science pre-service teacher education discussed above, the standard pre-service teacher school practicum has not changed in more than 100 years (Bacharach et al., 2007). Typically, a pre-service teacher observes science teaching for a short period, and then takes over teaching classes on their own. Usually, the student is placed with an experienced, cooperating teacher who acts as a mentor during the teaching experience. Student teaching is akin to an apprenticeship, with the pre-service teacher learning from the experienced teacher with an 'on-the-job' training approach. In this model, student teaching has a power structure and hierarchy that places a student teacher in a subservient position to a cooperating teacher, although the student teacher may have more recent knowledge of the field both in content and pedagogy such as science content knowledge and their perspectives on learning theories, assessment practices, and curriculum may be more consistent with the needs of twenty-first-century education than the cooperating teacher.

Some of the problems experienced by science pre-service teachers during school practicum are as follows:

- Anxiety during field and student teaching experiences, particularly the fear of failure.
- Anxiety inability to retain knowledge of ways to teach science effectively in the classroom that they learned in college.
- Inability to transform their scientific knowledge into PCK in both (primary elementary) and secondary schools.
- Handling the complexities involved with science teaching in school.
- The gap between theory and practice experienced by many pre-service (and cooperating) teachers.
- Under-resourcing of science learning and teaching in many schools, particularly with respect to technical staff.
- Use of digital technologies in science classrooms.
- Ineffective student learning because of inadequate pre-service teacher practice.
- Student disaffection with the learning and teaching of school science.
- Declining student attitudes towards school science.
- Mainstreaming students with special needs in school science labs.

A study by Menon and Sadler (2016) indicated that science content knowledge needs to be integrated with methods courses to enable pre-service teachers in practicing science teaching in some capacity, which might address some of the above issues. Other improvements in pre-service teacher preparation for practical science teaching include in-class cooperation between pre-service teacher and the cooperating teacher during the practicum. One of these models, which is being used more widely today, is coteaching.

Coteaching between pre-service teachers and in-service teachers during the school experience element of ITE programmes has been recognised as beneficial for student teachers, cooperating teachers (as in-school CPD), school students and teacher educators. A key difference between coteaching and other forms of collaborative instruction is that coteaching (without the hyphen) requires that coteachers support each other's learning as well as that of the school students. In the USA, the National Council for the Accreditation of Teacher Education (NCATE)'s Blue Ribbon panel on clinical preparation and partnerships noted the critical role of coteaching as a model for linking theory and practice in preparing teachers to teach (NCATE, 2010). Coteaching is an ideal methodology that teacher educators can use to simultaneously enhance their own reflective practice, as well as the reflective practice of pre- and in-service teachers. Coteaching also serves as pedagogy to improve school-based experience for pre-service and cooperating teachers. In coteaching, pre- and in-service teachers plan, teach and evaluate a series of lessons together.

The introduction of coteaching is a potential mitigating factor in relation to the attrition of new teachers from the profession (Murphy, 2016). García and Weiss (2019) discussed a high rate of teachers leaving the profession in the past five years. The figure is even higher in the UK, where almost 40% of new teachers leave teaching within a year of qualifying (Weale, 2015). Reasons for attrition include isolation, pressure to be 'outstanding', excessive home-based workload, and managing pupil misbehaviour. With coteaching, pre-service teachers can gain confidence rapidly by working alongside cooperating teachers, who can acquire both professional development and an expanded repertoire of teaching approaches and resources.

Coteaching provides a structure for pre- and in-service teachers to effect change via putting theory into practice and co-reflecting on how their teaching is developing towards 'ideal' practice. Coteaching is based firmly on the principle of sharing expertise. But just asking two or more teachers to work together is not always going to lead to successful coteaching. There needs to be a catalyst or spark which stimulates individual and/or collective 'aha' moments that inspire coteachers as they interact, and then to move towards developing successful coteaching partnerships. There also need to be workable structures to support the enactment of coteaching in school, for reflecting on practice, and for developing as coteachers. The rest of this chapter will explore these coteaching elements: the sparking of learning through coteaching and the structures for enacting, reflecting and developing coteaching practice.

### *10.3.1 Sparking Learning During Pre-service Teacher Teaching: Vygotskian Dramatic Collision*

Vygotsky developed a mechanism for *how* higher-order learning created between people is appropriated by individuals. His idea was termed '*kategoria*', a term used chiefly in Russian theatre and film, meaning a dramatic collision, which describes an inner tension causing a change in interest, motive or emotion and leads to change in behaviour. For Vygotsky, a dramatic collision must be experienced for the development of higher-order thinking, such as reflection. He argued that all that is taught is not always learned and does not necessarily lead to the development of HPFs, such as voluntary attention, reflection, and metacognition. Thus, in teaching, the idea of dramatic collision can be used to represent the 'sparks' that occur between teachers and their students which lead to learning as behavioural change (Murphy, 2016). In the classroom, pre-service teachers undergo many, many dramatic collisions, especially in the early days of teaching, as they 'find their feet'. The idea is that they manage these situations (which can be negative) and use them as learning opportunities.

Much deep learning comes from dramatic collisions that lead to self-examination. A further example of learning from self-examination as a result of dramatic collision comes from the reflections of Cristobal Carambo, who is an in-service teacher coteaching with a pre-service teacher:

> *Difference [between coteachers] achieves this [self-examination and change] because it does not allow for the reinforcement of the acceptable or the familiar, rather it provokes the examination of one's assumptions, and challenges our orthodox, habituated thoughts...the more difficult coteaching events forced me [Carambo] to re-examine my perspectives in light of those represented by my coteachers.* (Carambo & Stickney, 2009, p. 435).

### *10.3.2 ZPD in the Classroom*

How to share between coteachers learning can be theorised using Vygotsky's ZPD. The ZPD suggests conditions required for effective teaching and provides a set of tools that educators can apply to optimise the design and development of pre-service teacher teaching as an educational model for the crucial element of school experience in pre-service teacher education and as professional development for in-service teachers.

Vygotsky characterised the ZPD as "functions which have not yet matured but are in the process of maturing... 'buds' or 'flowers' of development rather than 'fruits' of development" (Vygotsky, 1978, p 86) and proposed that it represents "the domain of transitions that are accessible by the child" (Vygotsky, 1934/1987, p. 211). Many researchers describe the ZPD as interaction, which is collaboratively produced in the interaction between learner and teacher.

Vygotsky developed the ZPD within his CHT, which explains the basis of social transformation via the development of HPFs from the social to the individual. The ZPD was seen as a tool to promote development and learning. In ITE research and practice, the ZPD is used in this way to support pre-service teacher development. Interpretations of the ZPD suggest it is a two-way learning process with all participants learning through interactions with each other. Pre-service teachers, their cooperating teachers and students can expand their opportunities for learning while teaching.

**Table 10.2** Elements of Vygotsky's ZPD in the three teaching phases

| Teaching phase | Vygotskian elements |
|---|---|
| Planning | • Interaction between real and ideal form<br>• Buds of development |
| Practice | • Imitation<br>• Unity of affect and intellect |
| Evaluation | • Regression/recursion<br>• Structured reflection |

There are essential elements to the ZPD, without which development to the next stage might not occur at all, or just weakly. Murphy et al. (2014) selected those which they considered most appropriate for development as coteachers. They were interested in finding out which of these elements, if any, were particularly valuable to pre-service teachers in guiding their work. Murphy et al. (2014) assigned each of these elements to the three phases of planning, practice, and lesson evaluation. The elements ascribed to each phase are shown in Table 10.2:

### *10.3.3 Interaction Between Real and Ideal Form*

Vygotsky advocated the necessity from the earliest stages of development of having an ideal in mind because such an ideal provides the coteachers with motive and focus. The ideal is the perfect endpoint of development, which is never reached, but which provides a direction for the development. Yet, no development is possible without interaction between the ideal and real forms. If what is being moving towards is not known, they will never get there. Applying this element to teaching involves pre-service teachers identifying 'ideal' practice based on theories about learning and learners as they plan lessons. Pre-service teachers could seek and take advice from cooperating teachers regarding the practical elements they need to plan for, particularly in science lessons involving experiments and other practical activities. For example, this pre-service teacher reflected on how their planning had developed in terms of 'ideal' practice:

> *Before I focused on resources and how they worked, whereas after … I went: "okay this group didn't get this, and this is why I think they didn't get it, so this is what I'll do instead next time". It was much more detailed in terms of children's learning instead of the practical setup of the classroom.* (Pre-service teacher, interview, Murphy et al., 2014, p. 13)

### 10.3.4 Buds of Development

The best learning occurs within the ZPD when the learner is at a stage, a bud or flower according to Vygotsky (1978), which is proximal (or close) to the next level of development. Good teaching is not haphazard or spontaneous, but the result of planning, which requires the participation and involvement of all coteachers. Planning a ZPD involves pre-service teachers identifying their own 'buds' of development, as well as for the students (with the help of teacher educators and cooperating teachers) and the use of cultural tools to further develop these buds.

### 10.3.5 Imitation

Vygotsky's notion of 'imitation' is not copying but emulation (where pre-service teachers strive to equal or excel, not merely to copy) of a teaching activity as part of the learning process. Effective imitation within the ZPD pushes learning and development to a higher level, with successful emulation indicating the level of development of a maturing function. During practice, Vygotskian imitation can be enacted as a pre-service teacher observes and emulates the practice of an in-service teacher that is nearer to the ideal, thereby expanding their agency in relation to creating new practice.

> *...as such I seemed to move from the surface level to the more pedagogical and critical levels of reflection quicker, as I asked the questions the [observed] teacher would have asked, such as: 'where is the progression in this lesson?' 'Is this particular aspect of the lesson beneficial to learning?' 'How can you overcome the common misconceptions a child will make in this lesson?' etc.* (Pre-service teacher essay on reflection [Murphy et al., 2014, p.14]).

### 10.3.6 Unity of Affect and Intellect

The unity of affect and intellect in Vygotsky's ZPD suggests that emotion and learning are interdependent and foregrounds the importance of emotion in learning. Awareness of learning occurs via emotional experiences (negative as well as positive) and can be harnessed to develop better understanding their reactions as teachers. The pre-service teacher can use emotion to engage science students' interest and achievement.

### *10.3.7 Regression/Recursion*

Regression is key to deep learning. Please now refer back to the discussion of regression and recursion in the ZPD, which is covered earlier in this chapter (see Fig. 10.3). The concepts of regression and recursion are vital to pre-service and cooperating teachers in their own practice.

## 10.4 Developing Pre-service Science Teachers as Reflective Practioners

Coteaching provides ideal conditions for learning by creating a ZPD in which the collective achieves more than the individual (Murphy et al., 2015). The key characteristic underpinning coteaching is that pre-service teachers engage in discussions about practice and praxis with their cooperating teaching partners. It is this dynamic between participants which has been found to be key to making the often challenging practice of reflection more accessible, meaningful, and more rewarding. School-based work provides the potential for deep learning through reflection by pre-service teachers, who frequently use a 'trial and error' method. Pre-service teachers can be introduced to reflection 'tools', including Lampert-Shepel's (2008) model (see Fig. 10.4) of the reflective process.

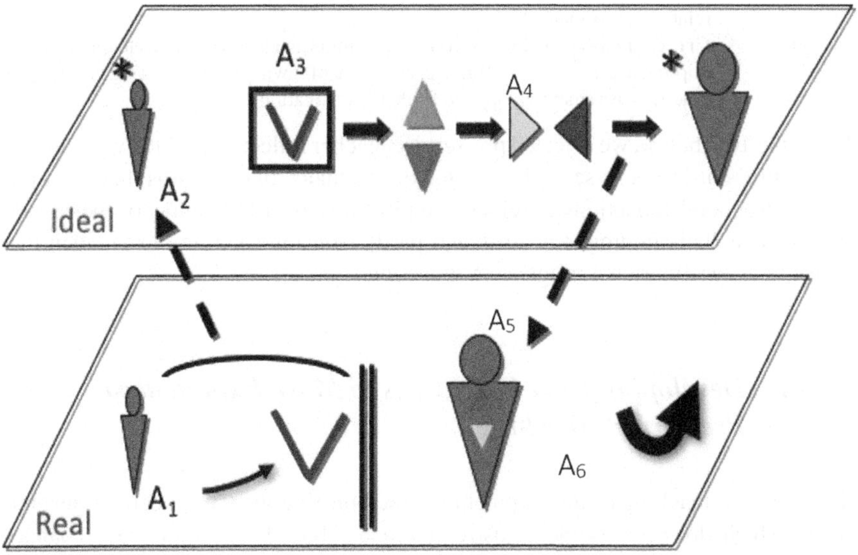


**Fig. 10.4** Reflection cycle. (Adapted from Lampert-Shepel, 2008)
KEY: *A1* agent of activity, *V* developmental gap, *A2* reflective stop, *A3* analysis of situation and choice of mediational means (represented as coloured triangles), *A4* modeling; *A5* transformation of model into practical action, *A6* reflexive control over the performance of a new practical action

The diagram (Fig. 10.4) presents an adapted version of Lampert-Shepel's (2008) model of reflection. The pre-service teacher (blue shape at phase A1) has experienced a problem in class (the 'real' plane in the diagram) which they cannot solve (represented as a developmental gap, V on the diagram). To address the problem, the pre-service teacher reflects outside the room, after the class in a different space, represented as the 'ideal plane' in the diagram (blue shape at phase A2 with asterisk, to show activity in the ideal plane on the diagram). Phase A3 represents the start of reflecting on the problem, and A4 comprises 'tools' that are used (such as other teachers, books, notes, computer sources, including the internet, etc.) and modelled into a strategy to address the problem, represented as A5. The pre-service teacher then goes back into the classroom, having internalised and ready to use the attempted solution, A6. If the problem is solved, then this cycle will be used again for subsequent reflective practice. If not, the pre-service teacher repeats the cycle from A1-A6, using more or different tools to develop and utilise a new strategy, until the issue is resolved.

The level of deep reflection described above, using Lampert-Shepel's (2008) model, can be considered as a working tool to achieve the highest level of reflection in an adapted model of Larrivee's (2008) tool which shows how pre-service teachers can move from the shallowest, and not very useful, to the highest level, which can lead to transformational change and improvement in pre-service teacher practice (Murphy et al., 2014). These levels are:

**Level 1:** **Surface reflection** (e.g., using evidence and making adjustments based on experience only).

**Level 2:** **Pedagogical reflection** (e.g., adjust methods and practices based on students' relative performance).

**Level 3:** **Critical pedagogical reflection** (e.g., commitment to continuous learning and improved practice; constructive criticism of own practice; seeing teaching practices as remaining open to further investigation).

For Table 10.3 below, we selected pre-service teacher reflections which were applied to the different levels as seen above. One of our major findings was that most pre-service teachers' reflections developed significantly over the course of their teaching practice. This finding was attributed partly to using the three levels identified above to assess their own progress in reflecting.

### *10.4.1 Developing from Early Days to More Experienced Pre-service Teaching*

The focus on teaching as development is based on Vygotsky's concept of development, which does not include "just evolutionary but also revolutionary changes, regression, gaps, zigzags, and conflicts" (Vygotsky, 1931/1997, p. 221). Such a complex, exciting, and visceral idea of development enables coteachers to remain confident, particularly in the early stages, when coteaching is not as straightforward.

**Table 10.3** Examples of pre-service teacher reflection at different levels

| Major category | Finding | Indicative quotes |
|---|---|---|
| Reflections on teaching with in-service teacher help (coteaching). | Identified huge benefits of coteaching, including working with a critical friend… | "I wouldn't have touched investigative science side before with a barge pole… Not on your life would I have given them [6-7 years] cups of water... too nervous of what they would do... my [coteaching] experience totally changed that." **Level 1** [surface reflection] (e.g., using evidence and making adjustments based on experience only). |
| Reflections on pre-service teacher solo teaching | Most pre-service teachers indicated that they progressed from evaluating resources and classroom activities to reflecting on children's learning. | "The content of reflection changed. Before I focused on resources… whereas after coteaching I went: ‚OK this group didn't get this and this is why I think they didn't get it, so this is what I'll do instead next time.' It was much more detailed in terms of children's learning..." **Level 2** [pedagogical reflection] (e.g., adjusts methods and practices based on students' relative performance). |
| Theory/practice | Direct and deep reflection on theory into practice and sometimes developing new theory from practice | "I have developed my reflective practice through the levels of progression… Reflection is arguably a process, not a method. It must be developed throughout...career. This journey...facilitating lessons which site pupils' learning in the forefront has begun and it will be interesting to chart the progress and effectiveness of my reflection throughout...[my] career". **Level 3** [critical pedagogical reflection] (e.g., commitment to continuous learning and improved practice; constructive criticism of own practice; seeing teaching practices as remaining open to further investigation). |

From Murphy et al. (2014)

The theoretical model for developing coteaching, based on Vygotsky's CHT, provides a framework for how higher-level pedagogical cognition develops during the activity of teaching. Vygotsky argued that high-level thinking for a sizeable proportion of people is required for social transformation. In pre-service teaching, the aim is to improve learning for all participants, resulting in developing the potential to transform classrooms to become more democratic, collaborative and focused on learning.

For such transformation to be realised, Stetsenko (2008) proposed that a *transformative-activist stance* needs to be adopted, which acknowledges teachers as *active agents* who effect change. In teaching, the potential for transformation is increased as pre-service teachers develop from early stages where they start as participants in the process, to the later stage where they are consciously sharing their contributions as teachers (Murphy & Carlisle, 2008).

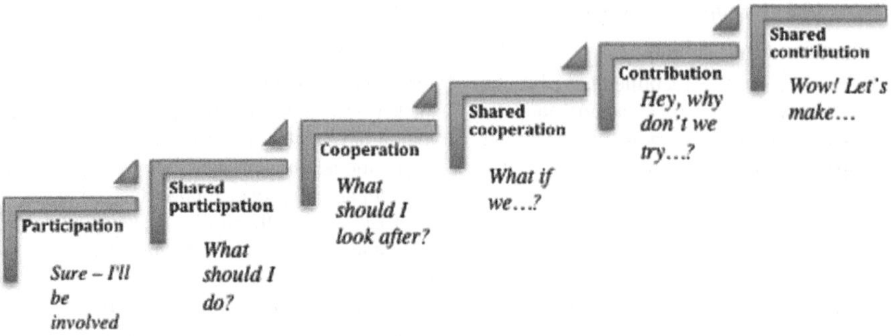


**Fig. 10.5** Stepwise development of pre-service-teacher collaboration in teaching

The model indicates progression through six stages, from teaching as initial *participation* in the process, whereby pre-service teachers focus mainly on their individual contexts (see Fig. 10.5).

Pre-service teachers are strongly advised to collaborate with cooperating and mentor in-service teachers. At this stage, the pre-service teacher might have a more theoretical, research-based conception of classroom teaching whereas the in-service teachers' focus is initially the context of the here and now in the specific classroom. Recognising and bringing these together characterises the next stage of *active participation*, which can lead to the third, *cooperation* stage, as each is developing areas of their own PCK, as described by Shulman (1986). Developing joint PCK represents the fourth, *shared cooperation* phase. By now, pre-service teachers will have developed the realisation that like any development, teaching does not improve in a 'straight line' and that it takes time. The latter stages are reached when the pre-service teacher is proactive in teaching by contributing their expertise to developing new materials/resources which can improve students' learning.

## 10.4.2 Summary of Vygotskian Ideas Applicable to Pre-service Teacher Development

Science pre-service teachers can be introduced to the constructivist notion that scientific knowledge is constructed, usually via experimentation, measurement, and observation by scientists in particular contexts. Science teachers, via their PCK, mediate this knowledge to students by 'translating' it into a form which is relevant and interesting to students. The students' prior knowledge of science is the starting point for the development of scientific concepts. Students, and all learners, need to interact with others to learn effectively. This interaction does not need to be direct – indirect interactions by way of books, computers, etc., are all authored by people.

The student is an active constructor of knowledge. Scientific concept development occurs in a cultural context that influences the form it takes (cultural mediation). Most of a student's scientific development evolves from social interactions, which, in turn, occur within, and are influenced by, specific cultural-historical settings. The ZPD enables learners' development from one stage to the next. The best learning takes place within the ZPD – and ZPDs can be created or co-created. As was covered in the previous section, it is important that pre-service teachers understand that scientific concept formation, from everyday to abstract, is not linear, but dialectical.

School-based learning can pose many challenges for pre-service teachers, especially in science. Taking over the class completely can leave a pre-service teacher floundering and losing confidence. Vygotsky's ZPD provides a construct which helps teacher educators, cooperating and mentor teachers to create conditions in which a pre-service teacher can be supported in their development as a successful new teacher. Elements of the ZPD which have been found most useful by pre-service teachers are the unity of real and ideal form, Vygotskian imitation, the unity of affect and intellect, and structured reflection. In essence, the unity of real and ideal form helps pre-service teachers to plan 'ideal' lessons, so that they are projecting towards the best learning for students, as opposed to planning resources and activities *per se.*

Vygotskian imitation helps pre-service teachers when observing, particularly observation of teachers who work well with students with whom the pre-service teacher experiences difficulties. Recording and emulating such behaviours in a way which is conducive to their own personalities (not blind copying) can enhance their skills in behaviour management. The unity of affect and intellect stresses the importance of emotion in learning for both students and teachers and indicates that some of the deepest learning develops from situations which cause strong emotions, both negative and positive. The structured reflection considered earlier gives pre-service teachers tools to reflect at higher levels, which allow them to create strategies to address issues and promote the development of becoming a reflective practitioner.

Most of the theory and practice development that has been gained from coteaching research is applicable to solo practice as well, as pre-service teachers teach alone during their school based ITE experiences. More about science coteaching can be found in books by Murphy and Scantlebury (2010) and Murphy (2016), as well as numerous articles on coteaching science in the research literature.

## 10.5 Vygotsky and in-Service Science Teacher Education

> *Science is an ongoing process. It never ends. There is no single ultimate truth to be achieved, after which all the scientists can retire. And because this is so, the world is far more interesting, both for the scientists and for the millions of people in every nation who, while not professional scientists, are deeply interested in the methods and findings of science.* (Sagan, 1980, p. xix).

In-service science teacher education is necessary for effectively implementing science teaching, which is relevant, engaging, linked more closely with contemporary science developments. The introduction to this chapter summarises how science teaching has changed historically in response to societal demands of the time.

Having completed a PhD in biology, followed by initial, induction and professional development as a science teacher, I can claim that my perception of science has changed significantly over the decades. I now consider science more as suggested by Sagan (1980) above. As a science teacher educator and researcher, it appears to me that change in science teaching is difficult in practice for teachers, who are being encouraged to use different ways from those they were taught to teach science, especially if they have been teaching for more than 10 years.

An international study of science teacher professional development research by Aldahmash et al. (2019) showed that the twenty-first-century focus of professional development of science teachers concentrated on PCK and constructivist learning and teaching approaches, including inquiry, problem-solving, learning cycles, and critical thinking skills. Fewer articles addressed lesson study, action research and STEM. They found that good-quality and effective professional development programmes for in-service science teachers are particularly important when initiating reforms in teaching and learning, and should be continuous and conducted as in-school-based inquiry activities, with follow up by way of a continuous programme of career-long professional development, building on previously acquired knowledge and skills.

Professional development of pre-service science teachers who are still in the preparation stage could also be researched, identifying different experiences of pre-service and in-service teachers. Research with pre- and in-service teachers coteaching science together, by sharing expertise, holds high potential for both parties. The pre-service teacher brings their learning of new science teaching developments and associated resources, whilst the in-service teacher acts as both colleague and mentor to support the pre-service teacher in school.

### *10.5.1 Addressing Challenges in Science Teacher Continuing Professional Development (CPD)*

The major challenge to providing excellent CPD arises when science teachers at all levels are faced with changing their usual practice to implement reforms of learning and teaching science. There is potential for strong emotional challenges as teachers may feel undermined, unequipped to teach in different ways, and not confident to apply the new learning in the classroom.

Most CPD takes place outside the school. Even if the sessions have been inspiring and very well facilitated, once a teacher gets back into class, they face their normal essential duties and find it very difficult time- and resource-wise to start teaching science differently. Two key Vygotskian concepts which are most helpful in preparing and enacting CPD programmes for science teachers are the *unity of affect and intellect*, and the ZPD.

## 10.5.2 *Unity of Affect and Intellect and the ZPD in Science Teacher CPD*

The unity of affect and intellect concept has been referred to in most chapters, and earlier in this chapter. In Vygotsky's words:

> *Affect and intellect are not two mutually exclusive poles, but two mental functions, closely connected with each other and inseparable, that appear at each age as an undifferentiated unity although they contain ever newer relations between affective and intellectual functions.* (Vygotsky, 1998, p. 239).

In other words, learning depends upon emotional engagement. It is easy to 'turn off' from learning when faced with boring, irrelevant, or disagreeable content and/or delivery. It is important, therefore, to create contexts for learning in science teacher CPD programmes which are aimed at engaging the teachers fully.

The unity of affect and intellect expands the ZPD, such that the ZPD for learning addresses both the affective and intellectual elements to ensure that teachers are highly engaged. In my experience, the best way to enable very good CPD is for the teachers to participate in all stages of the preparation, facilitation, and implementation of the programme. The cultural-historical context of the programme will influence the development of emotions which are needed to mediate successful establishment and maintenance of the ZPD.

Effective science teacher CPD depends upon emotionally positive collaboration and cooperation between teachers and CPD providers in a caring and nurturing environment. The individual emotional experience of each participant can be described as '*perezhivanie*' (see Chap. 3, Sect. 3.2) which is part of personal development and foundational to the continuing development of our HPFs (Chap. 3, Sect. 3.6). Sadly, the affective component of the ZPD is frequently ignored by Western psychologists (Levykh, 2008). Some ideas Levykh (2008) are included in the next section, to optimise a CPD Vygotskian framework:

## 10.5.3 *Optimising Science Teacher CPD Using a Vygotskian Framework*

- First, there is a need to focus on CPD as a *social* activity.
- Apply social constructivist approaches.
- Provide a caring and nurturing environment which utilises an expanded ZPD to embrace and enhance cooperation and collaboration in specific activities.
- Structure this environment so that learners feel comfortable expressing individual and social concerns, which can be addressed appropriately.
- Facilitate positive learning experiences and an appreciation for the subject.
- Facilitate creative risk-taking behaviour, together with acceptance of constructive criticism.

- Inspire trust and creativity for participants to internalise new knowledge and behaviours, which they can confidently externalise in their own teaching.
- CPD programmes should aim at transforming teachers, who can then transform their own science students.

## 10.6 Summary and Conclusion

Today, typing the phrase "Vygotsky and science teacher education" into Google yielded more than four million results. This chapter has revealed that the Vygotskian ideas most pertinent to science teacher education research are two-fold, in that science teachers are developing both themselves as teachers, and in school, where they are supporting students' learning in their science classes.

In terms of their own development, classroom management seems to be the greatest challenge for many science teachers (particularly pre-service and new teachers). ZPD creation by teacher educators and CPD providers can include facilitating workshops in which space is provided for teachers to think 'higher', by, for example, role-playing different scenarios which may occur in a school science lab. Swapping roles supports their ability to consider the situation from different perspectives. Also, ZPD elements such as the unity of affect and intellect, and regression/recursion, can be reflected on deeply to support difficult challenges when pre-service teachers are teaching during school placements, as they navigate and manage their emotional responses, and hold the awareness that their development will not be linear, but more like the tide, as ebb and flow.

Pre-service teachers can study and practice Vygotskian ideas and how they are integrated into science teaching and learning in terms of helping students to develop scientific concepts. This knowledge can support pre-service teachers' ability to prepare a classroom learning environment which enables students to use concepts, as opposed to simply defining them. It is also valuable for science pre-service teachers to discuss some of the current ideas about how we learn, such that students are aware of ways that they can monitor and take responsibility for their own learning.

The practice of coteaching in science pre-service teacher education is increasing, as significant benefits have been identified (Murphy, 2016). It has been described in this chapter and analysed using CHAT.

Pre-service teachers and in-service teachers will also learn a lot from the Nugent Study in Chap. 7 of this book, which describes the experience of an in-service teacher who, over a period of three to four years, worked on changing her traditional practice of chemistry teaching to using a Vygotskian social constructivist approach, with some surprisingly positive results.

Science teacher education in CPD and ITE programmes can be enhanced significantly by using social constructivist approaches. Such approaches are also recommended for school science teaching, so it is important to provide teachers with the appropriate tools work with in this regard.

## References


Aldahmash, A. H., Alshamrani, S. M., Alshaya, F. S., & Alsarrani, N. A. (2019). Research trends in in-service science teacher professional development from 2012 to 2016. *International Journal of Instruction, 12*(2), 163–178.

Bacharach, N., Heck, T., & Dahlberg, K. (2007). *Collaboratively researching the impact of a coteaching model of student teaching*. Paper presented at the annual meeting of the American Educational Research Association, Chicago.

Carambo, C., & Stickney, C. T. (2009). Coteaching praxis and professional service: Facilitating the transition of beliefs and practices. *Cultural Studies of Science Education, 4*, 433–441. Retrieved from https://doi.org/10.1007/s11422-008-9148-3

Chaiklin, S. (2003). The zone of proximal development in Vygotsky's analysis of learning and instruction. In A. Kozulin, B. Gindis, V. S. Ageyev & S. M. Miller (Eds.), *Vygotsky's educational theory in cultural context* (pp. 39–64). Cambridge University Press. Retrieved from https://doi.org/10.1017/CBO9780511840975

Dewey, J. (1916). *Democracy and education*. Macmillan.

García, E., & Weiss, E. (2019). *The teacher shortage is real, large and growing, and worse than we thought: The first report in the "perfect storm in the teacher labor market" series*. Economic Policy Institute.

Gredler, M., & Claytor Sheilds, C. (2008). *Vygotsky's legacy: A foundation for research and practice*. Guildford.

Kidman, G., & Fensham, P. J. (2020). Intended, achieved and unachieved values of science education: A historical review. In *Values in science education* (pp. 173–190).

Lampert-Shepel, E. (2008). Cultural-historical psychology and activity theory (CHAT) and case study design. *International Journal of Case Study Research and Application, 20*(2), 211–227.

Larrivee, B. (2008). Development of a tool to assess teachers' level of reflective practice. *Reflective Practice, 9*, 341–360. Retrieved from https://doi.org/10.1080/14623940802207451

Levykh, M. G. (2008). The affective establishment and maintenance of Vygotsky's zone of proximal development. *Educational Theory, 58*, 83–101.

Menon, D., & Sadler, T. D. (2016). *Changes in pre-service elementary teachers' science self-efficacy beliefs and its relation to science conceptual understandings in a science content course*. Paper presented at National Association for Research on Science Teaching.

Murphy, C. (2015). Reconceptualising the learning and teaching of scientific concepts. In D. Corrigan, C. Buntting, J. Dillon, A. Jones & R. Gunstone (Eds.) *The future in learning science: What's in it for the learner?* (pp. 127–150). Springer. Retrieved from https://doi.org/10.1007/978-3-319-16543-1

Murphy, C. (2016). *Coteaching in teacher education*. Critical Publishing Ltd.

Murphy, C., & Carlisle, K. (2008). Situating relational ontology and transformative activist stance within the "everyday" practice of coteaching and cogenerative dialogue. *Cultural Studies in Science Education, 3*, 493–506.

Murphy, C., & Scantlebury, K. (Eds.). (2010). *Coteaching in international contexts: Research and practice* (pp. 11–34). Springer. Retrieved from https://doi.org/10.1007/978-90-481-3707-7_2

Murphy, C., McCullagh, J., & Doherty, A. (2014). *Piloting a model for coteaching*. AERA. Retrieved from http://www.tara.tcd.ie/handle/2262/72312

National Council for Accreditation of Teacher Education (NCATE). (2010). *Transforming teacher education through clinical practice: A national strategy to prepare effective teachers*. NCATE.

Roth, W.-M., & Tobin, K. G. (Eds.). (2006). *Teaching together, learning together*. Peter Lang.

Sagan, C., Druyan, A., & Soter, S. (1980). *Who speaks for earth?* KCET and Carl Sagan Productions in association with BBC and Polytel International.

Shulman, L. S. (1986). Those who understand: Knowledge growth in teaching. *Educational Researcher, 5*(2), 4–14.

Stetsenko, A. (2008). From relational ontology to transformative activist stance in conceptualizing development and learning: Expanding Vygotsky's (CHAT) project. *Cultural Studies of Science Education, 3*, 471–491.

Tharp, R. G., & Gallimore, R. (1988). *Rousing minds to life: Teaching, learning, and schooling in social context*. Cambridge University Press.

Vygotsky, L. S. (1931/1997). The history of the development of higher mental functions. In *The collected works of L. S. Vygotsky, vol. 4* (pp. 1–251) (R. W. Reiber, Ed.; M. J. Hall, Trans.). Plenum.

Vygotsky, L. S. (1934/1987). Thinking and speech. In *The collected works of L. S. Vygotsky, vol. 1* (pp. 37–285). Edited by R. W. Reiber; translated by M. J. Hall. New York: Plenum.

Vygotsky, L. S. (1978). *Mind in society: The development of higher psychological processes* (M. Cole, V. John-Steiner, S. Scribner, & E. Souberman, Eds.). : Harvard University Press.

Vygotsky, L. S. (1987). Thinking and speech. In *The collected works of L. S. Vygotsky, vol. 1* (pp. 37–285) (R. W. Reiber, Ed.; M. J. Hall, Trans.). Plenum.

Vygotsky, L. S. (1998). The crisis at age seven. In *The collected works of L. S. Vygotsky, vol. 5* (pp. 289–296) (R. W. Reiber, Ed.; M. J. Hall, Trans.). Plenum.

Weale, S. (2015). Four in ten new teachers quit within a year. *The Guardian*. Retrieved from http://www.theguardian.com/education/2015/mar/31/four-in-10-new-teachers-quit-within-a-year

Zebroski, J. T. (1994). *Thinking through theory: Vygotskian perspectives on the teaching of writing*. Boynton/Cook Publishers, Inc..

# Part IV
# Epilogue

# Chapter 11
# Epilogue

## 11.1 Introduction

The work of Vygotsky has been shown in this book to have increasing relevance to science education research and practice. This short chapter sets out a possible framework to illustrate how specific Vygotskian concepts and principles might apply in different science education contexts. It is difficult and often considered bad practice to separate Vygotsky's constructs from each other, as it can be argued that all are embraced within his CHT. However, since Vygotsky died young and his theory was incomplete, it is up to researchers and practitioners to interpret his work in ways that promote positive change. The next section suggests one way to address this by showing that some of the Vygotskian constructs considered in this book are applicable to all the science education contexts discussed. The framework also indicates that some Vygotskian constructs are best applied to specific contexts only.

## 11.2 A Framework for Vygotsky and Science Education

There are many references to Vygotskian constructs in the science education research literature. The following list highlights those which I have come across most frequently:

- Zone of proximal development (ZPD).
- Social constructivism (SC).
- The importance of play in science education (Play).
- The role of imagination in science education (IiSE).
- Cultural-historical theory (CHT).
- Cultural-historical activity theory (CHAT).
- Cultural mediation (CM).
- Dialectical concept development (DCD).


C. Murphy, *Vygotsky and Science Education*,
https://doi.org/10.1007/978-3-031-05244-6_11

- Social situation of development (SSD).

These constructs represent both those that are mainly considered as Vygotskian, for example, the ZPD and CHT. Some constructs, however, have been assigned to Vygotsky, such as social constructivism. A simple, yet comprehensive 'broad-brush' framework for how these constructs (using the abbreviations assigned above) apply in different science education contexts is illustrated in Table 11.1. All the construct areas are present in each cell, indicating that they are all part of one grand theory; shaded cells indicate that the specific construct(s) is/are more important to that context. Science education research uses all the constructs.

## 11.3 Science Education and Scientific Research

There is a dialectical relationship between science education and scientific research: each depends on the other, in achieving all that science has discovered about the world around us.

In the past, science was enacted almost as a hobby for those who could afford it. Public engagement in science was entertainment-based. Scientists put on shows to wow the public with strange inventions and discoveries. One of the most famous of these shows was put on in the Royal College in London, by Galvani's nephew, Aldini, who showed that electricity existed as a force within the body. The body was that of George Foster, who had died from hanging. Aldini attached probes to Foster's body and powered up the battery. During the following hours the crowd saw his jaw quivering, his facial muscles contorted, and his left eye opened. Aldini thought his experiment had failed, as his intention was to bring Foster back to life. The story spread far and wide, with more exaggerated descriptions of Foster's body writhing and his head spinning. One consequence of this show was Mary Shelley's famous book *Frankenstein*, published in 1818.

Today, science is rarely seen as entertainment. It is often considered negatively in its association with war weapons, such as the development and use of the atom and hydrogen bombs. Some aspects of the science of medicine give science a good name, which can be tarnished by medical research that goes wrong, such as the thalidomide disaster. There is a perception of the negative side of science in terms of real and potential impacts on biodiversity and environmental deterioration, and there is the unknown future to be considered. The climate crisis is already causing havoc in different parts of the world. Science alone cannot mitigate the crisis, as many of the deterrents cannot be implemented, due to non-cooperation of businesses, politicians, and the public.

During the time I have been writing and editing this book, the COVID-19 coronavirus pandemic has killed more than 5 million people over the world. Scientists have developed highly effective vaccines, helping to reduce the number of deaths significantly. In addition to the vaccines, the necessity for lockdowns, with their associated physical and mental health problems and lack of face-to-face contact,

**Table 11.1** Vygotskian framework for science education contexts

| Contents | Vygotskian Constructs | | | | | | | | |
|---|---|---|---|---|---|---|---|---|---|
| Science Education Research | ZPD | CHT | SC | IiSE | Play | CHAT | CM | DCD | SSD |
| Early Years | ZPD | CHT | SC | IiSE | Play | CHAT | CM | DCD | SSD |
| Primary Science | ZPD | CHT | SC | IiSE | Play | CHAT | CM | DCD | SSD |
| Secondary Science | ZPD | CHT | SC | IiSE | Play | CHAT | CM | DCD | SSD |
| Informal Science Learning | ZPD | CHT | SC | IiSE | Play | CHAT | CM | DCD | SSD |
| Higher Education | ZPD | CHT | SC | IiSE | Play | CHAT | CM | DCD | SSD |
| Science Teacher Education | ZPD | CHT | SC | IiSE | Play | CHAT | CM | DCD | SSD |

**Key**: *ZPD* Zone of Proximal Development, *CHT* Cultural-historical Theory, *SC* Social Constructivism, *IiSE* Imagination in Science Education, *Play* Play in Science Education, *CHAT* Cultural Historical Activity Theory, *CM* Cultural Mediation, *DCD* Dialectical Concept Development, *SSD* Social Situation of Development

have partially mitigated the transmission of the virus. However, when people have failed to be compliant with the restrictions, there have been surges in the virus transmission and infection. Other barriers have included an 'antivax movement' encouraging people to refuse vaccinations against COVID-19. Sometimes the restrictions have been reduced, when cases of the virus, and the number of deaths has fallen significantly, but this has also been problematic, as some people have risked their own and others' chances of becoming very ill with the virus, even dying from it. The pandemic, it is hoped, will become epidemics in different countries, and there will be a reduction in the transmission and virulence of the COVID-19 coronavirus.

Questions are asked about how science education relates to science and whether science education can be used to influence scientific research positively. It could be argued that the disconnect between science education and scientific research is too wide, particularly regarding schooling. Science lessons in many schools are driven by a content-laden curriculum and largely summative assessment. There is little connection to scientific research, and not much attention paid to socio-scientific issues. It appears that school science education provides students with a knowledge base, much of which comprises the pseudo concept level of understanding. There are very few resources for teaching current science methodologies. Some school students will continue science learning in higher education, which is also beset by traditional lecture and lab-manual based science pedagogy, as well as being disconnected from scientific research as it is carried out today.

## 11.4 Future of Vygotsky in Science Education

The immediate future of Vygotsky and science education is for the wider implementation of Vygotskian principles and ideas into science education practice to support the development of twenty-first century skills in science in schools, colleges, and universities. Perhaps a Vygotskian dialectic relationship between science education and scientific research can be created, in which science education includes more learning about science in society and how curricular learning can be connected more closely to local and global scientific research. On the other side, scientific research could publish articles that are aimed at school and university students, which indicate how their research relates to the various aspects of the science curricula. Education publishers could be involved to facilitate such a scheme, mediated by science communicators who could assist in translating some of the terminology specific to each research context.

Developing stronger interaction between science at school and science in society should ensure that, from an early age, children are oriented within the scientific endeavour via formal and non-formal science learning experiences. An exciting outcome of bringing science education and scientific research closer also creates a much-expanded ZPD for learning in both sectors. Using the ZPD element: *the unity of real and ideal form* could help identify the 'ideal' outcomes of connecting science education and scientific research. The helps to project the direction of change on an ideal trajectory. Vygotskian imitation could enable students to carry out experiments which mimic some of the current scientific research, by scientists working together with schoolteachers, such as in the SIAB project described in Chapter 8. And of course, there is the recognition that any of these developments will be hard work and will take time, just as is required for the successful development of scientific concepts.

Future research on Vygotsky and science education could look at other ways of linking it more closely to address the gap between science and society. A useful model to frame such research is the data, information, knowledge, wisdom (DIKW) pyramid which was mooted in the late twentieth century (Ackoff, 1989; Zeleny, 1987). The DIKW pyramid illustrates the prevalence of scientific data, of which only some is analysed sufficiently to be considered as information. Not all this information is turned into useful knowledge, and only a small fraction of knowledge can be used as wisdom (see Fig. 11.1).

Data is objective, value-free and discrete, comprising facts and symbols; most of it held in technologies. Analysed data in a meaningful way provides information. When information is applied in specific contexts, it can be considered as knowledge. Wisdom is only possible when the knowledge has been evaluated in society to determine its potential use. Dallmeier-Tiessen et al. (2014) identified possible barriers to attaining wisdom from data, as illustrated in Fig. 11.1, as:

- Individual contributor incentives.
- Availability of a sustainable preservation infrastructure.
- Trustworthiness of the data, data usability, pre-archive activities.
- Data discovery.
- Academic defensiveness.
- Finance.
- Subject anonymity and personal data confidentiality.
- Legislation/regulation.

**Fig. 11.1** DIKW pyramid. (Adapted from Rowley, 2007)

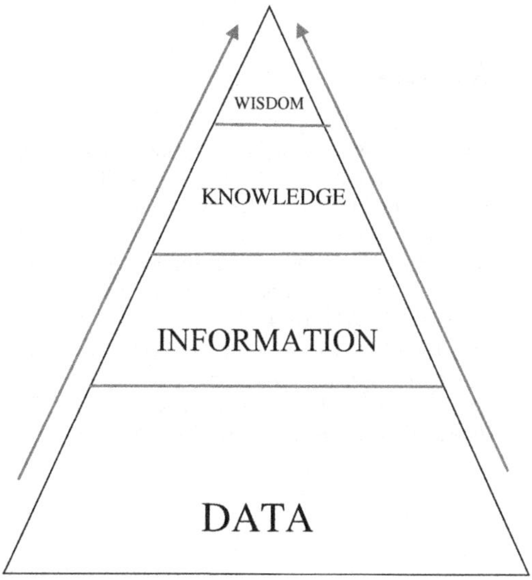

Other barriers could relate to the lack of sufficient interaction between scientists and society. An interesting study by Collins et al. (2018) reported findings which support the assessment that media representation of the issue of increasing antimicrobial resistance (AMR) tends to (wrongly) situate the responsibility for slowing this increase away from the individual. There is also fierce competition in science in terms of jobs, the 'publish or perish' culture and the need to get research articles into the leading journals. These journals fight to publish the most striking articles, running a risk for scientists to bury inconvenient data, or worse. Popular subjects are also more likely to be picked up, creating the need for scientists to focus on these areas, such as wine drinking, dementia, and artificial intelligence.

Science education should include critical studies of science communication. Students and school science curricula could mediate between science and society by ensuring that science education is relevant to both scientific research and its impact on society. Perhaps such an ideal could provide a new direction for science education mediated by Vygotskian principles in the various formal and informal science learning contexts described in this book.

During the COVID-19 pandemic, science has become more than usually prevalent in the public domain. Scientists have developed vaccines to prevent deaths and serious illness from this pandemic, but as I am writing there is a high degree of political wrangling regarding who gets the vaccine, and which type of vaccine. Considering the Vygotskian dialectic between science and society, had there been more high-level dialogue between scientists, health leaders, politicians, and other groups of global experts, to develop a workable strategy, we could have dealt much more effectively with reducing the transmission and infectivity of COVID-19 and all its mutants.

Hopefully, there might be positive outcomes for society in dealing with new pandemics and other human crises in the future.

## References

Ackoff, R. L. (1989). From data to wisdom. *Journal of Applied Systems Analysis, 16*(1), 3–9.

Collins, L. C., Jaspal, R., & Nerlich, B. (2018). Who or what has agency in the discussion of antimicrobial resistance in UK news media (2010–2015)? A transitivity analysis. *Health, 22*(6), 521–540.

Dallmeier-Tiessen, S., Darby, R., Gitmans, K., Lambert, S., Matthews, B., Mele, S., Suhonen, J., & Wilson, M. (2014). Enabling sharing and reuse of scientific data. *New Review of Information Networking, 19*(1), 16–43.

Rowley, J. (2007). The wisdom hierarchy: Representations of the DIKW hierarchy. *Journal of Information Science, 33*(2), 163–180.

Zeleny, M. (1987). Management support systems: Towards integrated knowledge management. *Human Systems Management, 7*(1), 59–70.

# Index




C. Murphy, *Vygotsky and Science Education*,
https://doi.org/10.1007/978-3-031-05244-6

The manufacturer's authorised representative in the EU is Springer
Nature Customer Service Centre GmbH, Europaplatz 3, 69115 Heidelberg,
Germany. If you have any concerns regarding our products, please
contact ProductSafety@springernature.com

Printed and bound by CPI Group (UK) Ltd, Croydon, CR0 4YY
30/09/2026
02225460-0013